Brain Dynamics

Springer
Berlin
Heidelberg
New York
Barcelona
Hong Kong
London
Milan
Paris
Tokyo

Springer Series in Synergetics

http://www.springer.de/phys/books/sssyn

SSSyn – An Interdisciplinary Series on Complex Systems

The success of the Springer Series in Synergetics has been made possible by the contributions of outstanding authors who presented their quite often pioneering results to the science community well beyond the borders of a special discipline. Indeed, interdisciplinarity is one of the main features of this series. But interdisciplinarity is not enough: The main goal is the search for common features of self-organizing systems in a great variety of seemingly quite different systems, or, still more precisely speaking, the search for general principles underlying the spontaneous formation of spatial, temporal or functional structures. The topics treated may be as diverse as lasers and fluids in physics, pattern formation in chemistry, morphogenesis in biology, brain functions in neurology or self-organization in a city. As is witnessed by several volumes, great attention is being paid to the pivotal interplay between deterministic and stochastic processes, as well as to the dialogue between theoreticians and experimentalists. All this has contributed to a remarkable cross-fertilization between disciplines and to a deeper understanding of complex systems. The timeliness and potential of such an approach are also mirrored – among other indicators – by numerous interdisciplinary workshops and conferences all over the world.

Hermann Haken

Brain Dynamics

Synchronization and Activity Patterns in Pulse-Coupled Neural Nets with Delays and Noise

With 82 Figures

Springer

BS

Professor Dr. Dr. h.c. mult. Hermann Haken
Institut für Theoretische Physik und Synergetik
Universität Stuttgart
Pfaffenwaldring 57/IV
70550 Stuttgart, Germany

Library of Congress Cataloging-in-Publication Data.

Haken, H.
Brain dynamics: synchronization and activity patterns in pulse-coupled neural nets with delays and noise / Hermann Haken.
p.cm. – (Springer series in synergetics, ISSN 0172-7389)
Includes bibliographical references.
ISBN 3540430768 (alk. paper)
1. Neural networks (Neurobiology) 2. Computational neuroscience. I. Title.
II. Springer series in synergetics (Unnumbered)
QP363.3.H354 2002 573.8'5–dc21 2002070797

ISSN 0172-7389

ISBN 3-540-43076-8 Springer-Verlag Berlin Heidelberg New York

Springer-Verlag Berlin Heidelberg New York
a member of BertelsmannSpringer Science+Business Media GmbH

http://www.springer.de

Data conversion by LE-TEX, Leipzig
Cover design: *design & production*, Heidelberg
Printed on acid-free paper SPIN: 10860232 55/3141/tr - 5 4 3 2 1 0

6|12|06

Foreword

Twenty-Five Years of Springer Series in Synergetics

The year 2002 marks the 25th anniversary of the Springer Series in Synergetics. It started in 1977 with my book "Synergetics. An Introduction. Nonequilibrium Phase Transitions and Self-Organization in Physics, Chemistry and Biology". In the near future, the 100th volume of this series will be published. Its success has been made possible by the contributions of outstanding authors who presented their quite often pioneering results to the science community well beyond the borders of a special discipline. Indeed, interdisciplinarity is one of the main features of this series. But interdisciplinarity is not enough: The main goal is the search for common features of self-organizing systems in a great variety of seemingly quite different systems, or, still more precisely speaking, the search for general principles underlying the spontaneous formation of spatial, temporal or functional structures. The objects studied may be as diverse as lasers and fluids in physics, pattern formation in chemistry, morphogenesis in biology, brain functions in neurology or self-organization in a city. As is witnessed by several volumes, great attention is being paid to the pivotal interplay between deterministic and stochastic processes, as well as to the dialogue between theoreticians and experimentalists. All this has contributed to a remarkable cross-fertilization between disciplines and to a deeper unterstanding of complex systems. The timeliness and potential of such an approach are also mirrored – among other indicators – by numerous interdisciplinary workshops and conferences all over the world.

An important goal of the Springer Series in Synergetics will be to retain its high scientific standard and its good readability across disciplines. The recently formed editorial board with its outstanding scientists will be a great help.

As editor of this series, I wish to thank all those who contributed to its success. There are the authors, but, perhaps less visibly though of great importance, the members of Springer-Verlag, who over the past 25 years indefatigably have taken care of this series, in particular Dr. Helmut Lotsch, Dr. Angela Lahee, Prof. Wolf Beiglböck and their teams.

Stuttgart, June 2002 *Hermann Haken*

Preface

Research on the human brain has become a truly interdisciplinary enterprise that no longer belongs to medicine, neurobiology and related fields alone. In fact, in our attempts to understand the functioning of the human brain, more and more concepts from physics, mathematics, computer science, mathematical biology and related fields are used. This list is by no means complete, but it reflects the aim of the present book. It will show how concepts and mathematical tools of these fields allow us to treat important aspects of the behavior of large networks of the building blocks of the brain, the neurons.

This book applies to graduate students, professors and researchers in the above-mentioned fields, whereby I aimed throughout at a pedagogical style. A basic knowledge of calculus should be sufficient. In view of the various backgrounds of the readers of my book, I wrote several introductory chapters. For those who have little or no knowledge of the basic facts of neurons that will be needed later I included two chapters. Readers from the field of neuroscience, but also from other disciplines, will find the chapter on mathematical concepts and tricks useful. It shows how to describe spiking neurons and contains material that cannot easily be found in conventional textbooks, e.g. on the handling of δ-functions. Noise in physical systems – and thus also in the brain – is inevitable. This is true for systems in thermal equilibrium, but still more so in active systems – and neuronal systems are indeed highly active. Therefore, I deal with the origin and effects of noise in such systems.

After these preparations, I will deal with large neural networks. A central issue is the spontaneous synchronization of the spiking of neurons. At least some authors consider it as a basic mechanism for the binding problem, where various features of a scene, that may even be processed in different parts of the brain, are composed to a unique perception. While this idea is not generally accepted, the problem of understanding the behavior of large nets, especially with respect to synchronization, is nevertheless a fundamental problem of contemporary research. For instance, synchronization among neurons seems to play a fundamental role in epileptic seizures and Parkinson's disease. Therefore, the main part of my book will be devoted to the synchronization problem and will expose various kinds of integrate and fire models as well as what I called the lighthouse model. My approach seems to be more realistic than conventional neural net models in that it takes into account the detailed dynamics of axons, synapses and dendrites, whereby I consider arbitrary couplings between neurons, delays and the effect of noise. Experts will notice that this approach goes considerably beyond those that have been published so far in the literature.

I will treat different kinds of synaptic (dendritic) responses, determine the synchronized (phase-locked) state for all models and the limits of its stability. The role of non-synchronized states in associative memory will also be elucidated. To draw a more complete picture of present-day approaches to phase-locking and synchronization, I present also other phase-locking mechanisms and their relation, for instance, to movement coordination. When we average our basic neural equations over pulses, we reobtain the by now well-known Wilson–Cowan equations for axonal spike rates as well as the coupled equations for dendritic currents and axonal rates as derived by Nunez and extended by Jirsa and Haken. For the sake of completeness, I include a brief chapter on the equations describing a single neuron, i.e. on the Hodgkin–Huxley equations and generalizations thereof.

I had the opportunity of presenting my results in numerous plenary talks or lectures at international conferences and summer schools and could profit from the discussions. My thanks go, in particular, to Fanji Gu, Y. Kuramoto, H. Liljenström, P. McClintock, S. Nara, X.L. Qi, M. Robnik, H. Saido, I. Tsuda, M. Tsukada, and Yunjiu Wang. I hope that the readers of my book will find it enjoyable and useful as did the audience of my lectures. My book may be considered complementary to my former book on "Principles of Brain Functioning". Whereas in that book the global aspects of brain functioning are elaborated using the interdisciplinary approach of synergetics, the present one starts from the neuronal level and studies modern and important aspects of neural networks. The other end is covered by Hugh R. Wilson's book on "Spikes, Decisions and Actions" that deals with the single neuron and the action of a few of them. While his book provides readers from neuroscience with an excellent introduction to the mathematics of nonlinear dynamics, my earlier book "Synergetics. An Introduction" serves a similar purpose for mathematicians and physicists.

The tireless help of my secretary Ms. I. Möller has been pivotal for me in bringing this book to a good end. When typing the text and composing the formulas she – once again – performed the miracle of combining great speed with utmost accuracy. Most of the figures were drawn by Ms. Karin Hahn. Many thanks to her for her perfect work.

Last but not least I thank the team at Springer-Verlag for their traditionally excellent cooperation, in particular Prof. W. Beiglböck, Ms. S. Lehr and Ms. B. Reichel-Mayer.

Stuttgart, June 2002 *Hermann Haken*

Contents

* Sections marked by an asterisk are somewhat more involved and can be skipped.

Part II. Spiking in Neural Nets

Part I

Basic Experimental Facts and Theoretical Tools

1. Introduction

1.1 Goal

The human brain is the most complex system we know of. It consists of about 100 billion neurons that interact in a highly complicated fashion with each other. In my book I will conceive the brain as a physical system and study the behavior of large neural nets. Neurons are nonlinear elements. Most of them are able to produce trains of individual spikes, by which information between the neurons is exchanged. In addition, it is by now generally believed that correlations between spike trains play an important role in brain activity. One particular experimentally observed phenomenon is that of synchronization between the "firing" of neurons, where Fig. 1.1 shows an idealized case. A number of authors (see references) believe that synchronization is a fundamental mechanism that allows us to understand how the brain solves the binding problem. For instance, a lemon may be characterized by its shape, colour, smell, its name in various languages, and so on. Though all these aspects are processed in distinct parts of the brain, we nevertheless conceive the lemon as an entity. Synchronization may also help to identify individual parts of a scene as belonging to the same object. It must be noted, however, that these interpretations of the significance of synchronization are subject to ongoing critical discussions. On the other hand, synchronization among large groups of neurons may also be detrimental to healthy behavior. For instance, Parkinsonian tremor and epileptic seizures are believed to be caused by such a mechanism. At any rate, understanding synchronization and desynchronization are fundamental problems in modern brain research.

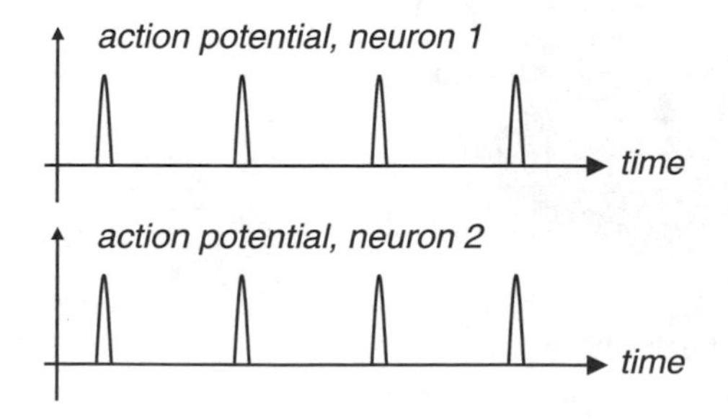

Fig. 1.1. Synchrony between two spike trains (schematic). For more details cf. Sects. 1.1 and 1.3

Studying networks of neurons means that we pick a specific level of investigation. In fact, each neuron is a complex system by itself, which at the microscopic level has a complicated structure and in which numerous complex chemical and electrochemical processes go on. Nevertheless, in order to model the behavior of a neural net, in general it is possible to treat the behavior of an individual neuron using a few characteristic features. The reason lies in the different time and length scales of the various activities, a fact that has found its detailed theoretical justification in the field of synergetics. Beyond that, for many practical purposes, the selection of the relevant neuronal variables and their equations largely depends on the experience and skill of the modeler as well as on his/her ability to solve the resulting network equations. Clearly, when we go beyond neural nets, new qualitative features appear, such as perception, motor-control, and so on. These must always be kept in mind by the reader, and in my book I will point at some of the corresponding links.

1.2 Brain: Structure and Functioning. A Brief Reminder

A complete survey of what science nowadays knows about the brain would fill a library. Therefore it may suffice here to mention a few relevant aspects. The white-gray matter of the brain is arranged in the form of a walnut (Fig. 1.2). As has been known for some time, through the effects of injuries or strokes,

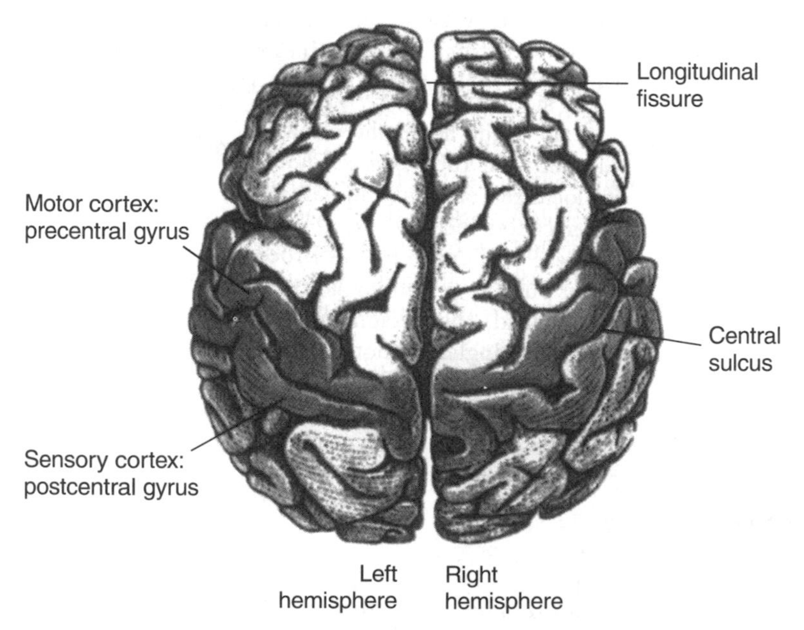

Fig. 1.2. The brain seen from above

there are localized areas in the brain that can be considered as centers for specific processes, such as tactile sensations, movement control, seeing, hearing, speech production, etc. These early findings by medical doctors could not only be substantiated, but also extended by modern physical methods, such as magnetoencephalograms, electroencephalograms, positron emission spectroscopy, magnetic resonance imaging, and so on. Since I described these approaches in my book "Principles of Brain Functioning", and since they may be found in other text books as well, I will not elaborate on these methods here. By means of these methods, it has become clear, however, that there are pronounced interconnections between the various regions of the brain, whereby learning and plasticity may play an important role. For instance, when a finger of the hand of a monkey is removed, the corresponding brain area shrinks and is largely taken over by the neuronal endings ("afferent nerve fibers") corresponding to neighbouring fingers. Thus the concept of localized areas must be taken with a grain of salt. As may transpire from what I have just said, it must be left open what we consider as part of a neural network. So in the following, taking a broad view, we may think of a neural network as one that is contained in an individual area, but also as one that comprises parts of different areas as well. I believe that here much has to be done in future experimenal and theoretical research.

After having said this, I may proceed to a preliminary discussion of individual network models.

1.3 Network Models

While the model of a single neuron is by now well established, being based on the fundamental work by Hodgkin and Huxley, modern theoretical work deals with the branching of the solutions of the Hodgkin–Huxley equations and their modifications and generalizations under the impact of external and internal parameters. In other words, an intense study of the bifurcations of the Hodgkin–Huxley equations and related equations is performed. When we proceed to two or few neurons, mostly computer models are invoked, including numbers of up to hundreds of neurons, whereby highly simplified dynamics must be used. Basically two kinds of couplings between neurons have been treated in the literature. One is the model of sinusoidal coupling, depending on the relative phase of two neurons. This theory is based on the concept of phase oscillators, i.e. on devices whose dynamics can be described by a single variable, the phase. Corresponding approaches have a long history in radio-engineering and later in laser physics, where the coupling between few oscillators is dealt with. The coupling between many biological or chemical phase oscillators has been treated in the pioneering works by Winfree and Kuramoto, respectively. An excellent survey of the development of this approach can be found in the article by Strogatz (see references). Applications to neural nets have been implemented by Kuramoto and others. More recent and more realistic approaches rest on the study of the interaction between neuronal spike trains. A simple but equally important model has been developed by Mirollo and Strogatz and further continued by Geisel and coworkers. This model was originally introduced by Peskin to explain the self-synchronization of the cardiac pacemaker. More recent work on this class of models, called integrate and fire models, has been performed by Bresloff, Coombes and other authors.

The central part of this book will be devoted to networks composed of many neurons coupled by spike trains. Hereby I first develop what I call the lighthouse model, which can be treated in great detail and rather simply and yet allows us at the same time to take into account many different effects including delays between neurons and noise. As we will see, under typical initial conditions a steady synchronized state evolves, whose stability and instability we will study in detail. Depending on the interactions between the neurons, i.e. depending on their synaptic strengths, a change of modus from long spike intervals to short spike intervals may happen. We allow for arbitrary couplings with a special constraint, however, that allows for synchronized states. We will elucidate the relation between the lighthouse model and integrate and fire models in detail, whereby we perform in both cases a rather complete stability analysis that goes far beyond what has been known so far in the literature.

We will also discuss the mechanisms of associative memory based on these models and include for the sake of completeness sinusoidal couplings at various levels of biological organisation, i.e. both at the neuronal level and that of limbs.

Finally, we will show how phase-averaged equations can be deduced from our basic equations, whereby we recover the fundamental equations of Wilson and Cowan as well as of Nunez, Jirsa and Haken. These equations have found widespread applications to the understanding of the formation of spatio-temporal activity patterns of neuronal nets. We illustrate the use of these equations in particular by means of the Kelso experiments on finger movements. This allows us to show how the present approach allows one to go from the individual neuronal level up to the macroscopic observable level of motion of limbs.

1.4 How We Will Proceed

We first give a descriptive outline on the structure and basic functions of an individual neuron. This will be followed by the presentation of typical and important effects of their cooperation, in particular the experimental evidence of their synchronization under specific conditions.

In Chap. 4 we will be concerned with theoretical concepts and mathematical tools. In particular we show how to represent spikes, what is meant by phase and how to determine it from experimental data. Furthermore we will show how the origin and effect of noise can be modelled.

Chapters 5 and 6 are devoted to the lighthouse model with its various aspects.

Chapter 7 provides a bridge between the lighthouse model and the integrate and fire models, where a broad view is taken.

In Chapter 8 we treat integrate and fire models of different kinds from a unifying point of view and explore in particular their stability and instability properties.

Chapter 9 is devoted to sinusoidal couplings and shows the usefulness of this kind of model by means of applications to neurons as well as to movement coordination.

As already mentioned, Chap. 10 deals with phase-averaged equations for axonal spike rates and dendritic currents, whereas Chap. 11 gives, for sake of completeness, an outline of Hodgkin–Huxley equations and related approaches, that means that this chapter deals with the individual neuronal level.

The book concludes with Chap. 12 “Conclusion and Outlook”.

2. The Neuron – Building Block of the Brain

2.1 Structure and Basic Functions

Though there are about 20 different types of neurons, their structure is basically the same. A neuron is composed of its soma, its dendrites that quite often form a treelike structure and the axon that, eventually, branches (Figs. 2.1 and 2.2). Information produced in other neurons is transferred to the neuron under consideration by means of localized contacts, the synapses, that are located on the dendrites and also on the cell body. Electrical charges

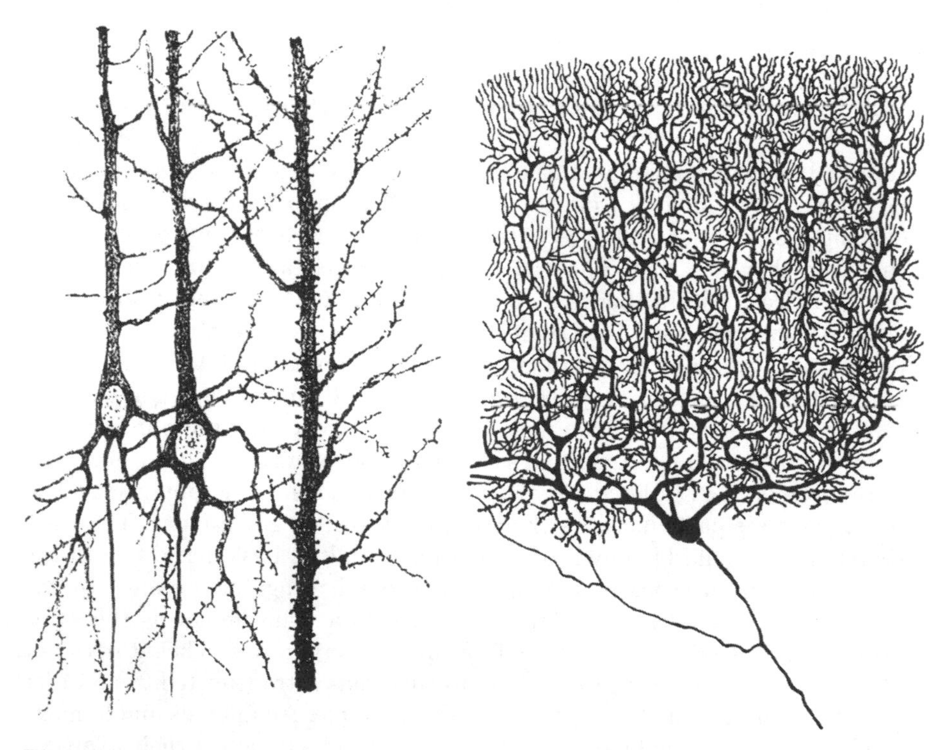

Fig. 2.1. Examples of neurons. L.h.s.: Pyramidal cell, r.h.s.: Purkinje cell (after Bullock et al., 1977)

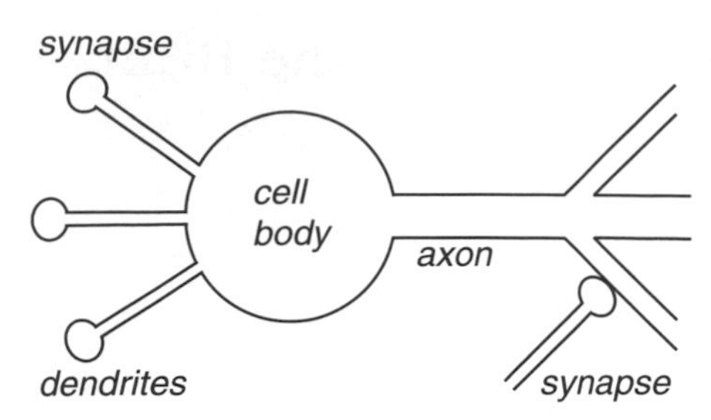

Fig. 2.2. Scheme of a neuron

produced at the synapses propagate to the soma and produce a net postsynaptic potential. If the postsynaptic potential at the soma is sufficiently large to exceed a threshold value, typically a depolarisation of 10–15 mV, the neuron generates a brief electrical pulse that is called a spike or action potential, at its axon hillock. The axon hillock is the point of connection between the soma and the axon. The spikes run down the axon, finally reach the synapses that, in a way to be discussed below, transfer the information to another neuron. In order to be able to model the functioning of a neuron, we have to deal with these processes in more detail. In this chapter we will be satisfied by a qualitative discussion with only a few mathematical hints.

2.2 Information Transmission in an Axon

Information transmission in an axon is based on electrical processes that are, however, rather different from those in metallic conductors involving electric currents. While in metals the carrriers of electric charge are electrons, in the axon they are ions. These electrically charged atoms are much heavier than electrons. Furthermore, a nerve fibre is much thinner than a conventional metallic wire. The diameter of an axon is only about 0.1–20 μm (1 μm = 10^{-9} m). The longitudinal resistance of an axon of 1 m length is as high as the resistance of a copper wire more than 10^{10} miles long. Quite cleary, electrical processes in an axon must be quite different from those in wires. In order to understand the mechanism of information transfer, measurements were made both in isolated nerve preparations and in living organisms. The squid possesses particularly thick axons, the so-called giant axons. They are, therefore, particularly suited for such studies and all basic insights into the function of the nervous system were first found using these axons. In the meantime we know that the kind of information transmission is the same both within the same organism and for different organisms. Thus it does not play a role, whether pain from a limb to the brain is transmitted or an order from the brain to a limb is transmitted, for example. All animals and humans use basically only one kind of information transmission along their axons.

Experimentally it can be shown that at a resting nerve fiber that does not transmit information a small electric potential between its inner and outer side is present. This potential is called the *resting potential.* The inner part of the nerve fibre is negatively charged as compared to the outer liquid in which the nerve fibre is embedded. This potential is about 70 mV. The reason for this resting potential is the unequal distribution of ions within and outside the axon and is due to special properties of the axon membrane, which has different permeabilities for different ions. An energy-consuming process, the sodium–potassium pump, maintains the unequal distribution of ions. What happens at an active neuron that transmits information? This can be understood by means of the following experiment (Fig. 2.3). If a small current is injected into the axon via an electrode (v1), at the position (v3) the resting potential is lowered, i.e. the potential difference between the inside and the outside is decreased. This is called depolarization. As can be expected from the electrical properties of the axon, this depolarization is only weakly registered at an electrode (v2) that is further away. If the current through the electrode (v1) is enhanced, the depolarization increases correspondingly. At a certain polarization (threshold), a new phenomenon appears. Suddenly a short reversal of charges occurs in a small area. In other words, for a short time the outer side of the axon becomes negative as compared to its inner side. Most remarkably, this change of potential is considerably larger than expected for the level of the injected current. Also the duration of the reversal of the potential is not influenced by the duration of the injected current pulse. Quite clearly, we are dealing with an active process of the axon. If the pulse at the electrode (v1) is increased further, the level and duration of this reaction will not change. Thus, we are speaking of an *all or nothing signal.* In other words, this signal does not occur at a subthreshold electric excitation, but fully occurs at a superthreshold excitation. This change of potential can be registered at a distant third electrode with the full level and with only a small delay. Thus, the response migrates further, and with increasing distance no decrease of the potential occurs. Clearly, this property is important in the transfer of information from one neuron to another. The short reversal of voltage is called a *nerve pulse* or *action potential.* Its duration is about one thousandth of a second. Quite often, it is called a *spike.*

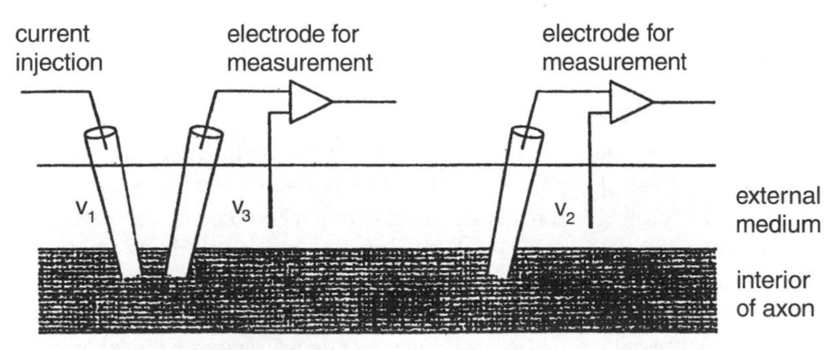

Fig. 2.3. Scheme of an experiment on the origin of a pulse

How can the formation of such a nerve pulse be explained? With depolarization, the membrane is electrically excited and ion channels open. Rapidly ions can migrate through these channels and thus cause the reversal of charge. But then the opening of other channels and thus the migration of other ions causes a decrease of this voltage reversal. The sodium–potassium pump maintains the unequal ionic distribution. A nerve pulse migrates in a nerve cell from the beginning of the axon, the axon hillock, in the direction of a synapse. Its velocity can be up to 100 m/s, corresponding to 360 km/h. In spite of the extremely high longitudinal resistance of an axon, the electric pulse can thus be transmitted via the axon extremely rapidly. This is made possible because the charge carriers, the ions, need not move along the axon, but perpendicularly through a very thin membrane. With modern methods (patch clamp experiments), it is possible to study even the processes at individual channels. We will not be concerned with these microscopic processes here. How are electrical excitations of axons produced in nature? Here, of course, no electrodes are introduced in the axon and no current will be injected artificially. In many cases, electric excitations stem from other nerve cells and are transferred via the synapses. Electric excitations originate in the sensory organs at receptors, which are special cells that transform external excitations into electrical excitations. For instance, light impinging on receptors in the retina is finally transformed into electric excitations that are then further transmitted.

2.3 Neural Code

How can information be transmitted by means of neural pulses? We have to remember that in a specific nerve fiber all nerve pulses have the same intensity and duration. Thus there is only one signal. In the nervous system, sequences of spikes are used, whose temporal distance, or, in other words, whose frequency, is variable. The stronger a nerve fiber is excited, the higher the frequency. Note that the meaning of a piece of information, whether for instance it is a piece of visual, acoustic or tactile information, cannot be encoded using the frequency of the nerve impulse. The meaning of a piece of impulse information in an organism is fixed by the origin and destination of its nerve fiber. This means for instance that all action potentials that are transmitted via nerve fibers stemming from the eye contain visual information. These nerve fibers finally lead, via several switching stations, to a special part of the brain, the visual cortex. The same is true for other nerve fibers. Also the quality of an excitation, for instance the color of an object, is determined by the kind of nerve fiber. For instance, separate fibers originate from different receptors for color in the eye. Sensory cells are specialized nerve cells that convert external excitations, such as light, temperature variations, sound, a.s.o. into electrical excitations. Sensory cells in a way are interpreters between the external world and the nervous system, but they react only quite specifically to specific excitations. For our later modelling, these observations are of fundamental importance, because they lie at the root of the possible universality of network models.

2.4 Synapses – The Local Contacts

At most synapses information transmission is not achieved by means of electrical pulses but by means of chemical substances, the so-called neurotransmitters. Figure 2.4 shows a highly simplified sketch of the structure of a chemical synapse. Between the two nerve cells there is a small gap across which information is transmitted by the migration of chemical substances. In detail, the following processes go on. When an action potential reaches the synapse, transmitters are released from small vesicles and proceed from there to the synaptic gap. Here they diffuse to the other (postsynaptic) side and dock on specific molecules, the receptors. The transmitter molecules fit to the receptors like a key in a lock. As soon as the transmitter substances dock at the receptors, this influences specific ion channels, causing a migration of ions and thus a depolarization of the membrane. The higher the frequency of the incoming action potentials (pulses), the more transmitter substance is released and the larger the depolarization on the postsynaptic side. The transmitter molecules are relatively quickly decomposed and the individual parts return to the presynaptic side. There they are reconstructed to complete transmitter molecules and stored in the vesicles. The now unoccupied receptors can again be occupied by new transmitter molecules. If at the presynaptic sides no action potentials arrive, no more transmitter molecules are liberated from the vesicles and the receptors remain unoccupied. Thus the depolarization decreases. The transmission of the excitation via the synapses leads to a local potential at the cell body. Only if this potential exceeds a certain threshold at the axon hillock are action potentials generated that are

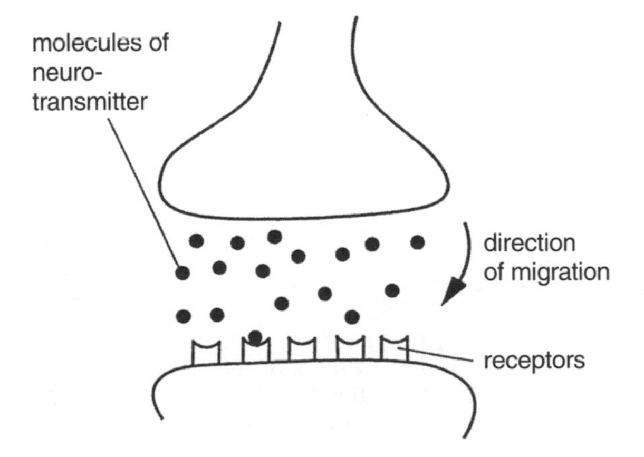

Fig. 2.4. Scheme of a synapse

then transmitted along the axon. In general, the transmission of excitation at a single synapse is not sufficient to produce a superthreshold depolarization. But nerve cells are connected with many other nerve cells, because on the dendrites and on the cell body of the nerve cell many synapses connecting with other nerve cells are located. The excitations which come in across all these synapses contribute to the local potential at the cell body. It is important to note that not all synapses are excitatory, but there are also inhibitory synapses that decrease the excitation of the local potential at the cell body. The actions of excitatory and inhibitory synapses are thus processed in the region of the cell body. As mentioned above, the corresponding nerve cell transmits nerve pulses across its axon only, when at the beginning of the axon, i.e. at the axon hillock, a superthreshold depolarization occurs. If the potential remains under this threshold, no nerve pulses will be carried on. The higher the superthreshold local potential, i.e. the higher the depolarization, the higher is the frequency with which the axon potentials are carried on from this nerve cell. Clearly, the threshold of the information transmission ensures that small random fluctuations at the cell body don't lead to information transmission via nerve pulses. The inhibitory synapses have an important function also, because they impede an extreme amplification of electric excitation in the nervous system.

2.5 Naka–Rushton Relation

For our models that we will formulate later, we need a quantitative relation between the stimulus intensity P that acts at the site of spike generation and the firing rate, i.e. the production rate of spikes. For quite a number of neurons this relation has a rather general form provided the stimulus intensity P is constant and we are considering the resulting steady state in which the firing rate is time-independent (Figs. 2.5 and 2.6).

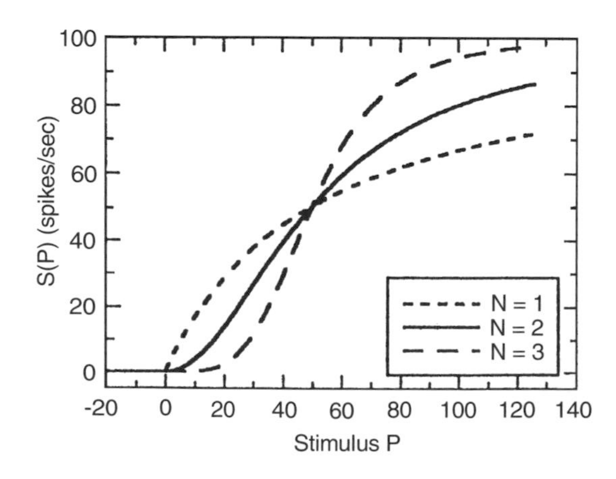

Fig. 2.5. The Naka–Rushton function for $N = 1, 2$ and 3 (after Wilson, 1999)

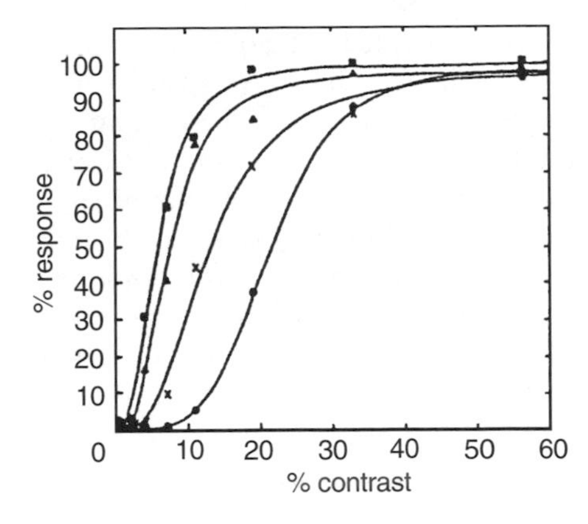

Fig. 2.6. Experimentally observed spike rates of four different neurons with fits of (2.1) (after Albrecht and Hamilton, 1982)

The relation is the Naka–Rushton formula

$$S(P) = \begin{cases} rP^N/(\Theta^N + P^N) & \text{for } P \geq 0\,, \\ 0 & \text{for } P < 0\,. \end{cases} \tag{2.1}$$

The meanings of the constants r and Θ become clear if we consider special cases. If

$$P \gg \Theta\,, \tag{2.2}$$

we obtain

$$S(P) \approx r\,, \tag{2.3}$$

so that r is the maximum spike rate. If we choose, however,

$$P = \Theta\,, \tag{2.4}$$

we obtain

$$S(P) = \frac{rP^N}{2P^N} = \frac{r}{2}\,, \tag{2.5}$$

i.e. (2.4) determines the point at which (2.1) reaches half its maximum. The exponent N is roughly a measure for the steepness of the curve $S(P)$. Typical values of N that match experimentally observed data range from 1.4 to 3.4. In the literature, a number of similar functions S are used in describing the spike rate. All of them have the following properties:

1. There is a threshold for P close to zero.
2. There is roughly a linear region in which

$$S(P) \propto P\,. \tag{2.6}$$

3. For large enough values (see (2.2)), S becomes constant, an effect called *saturation*.

For our later model we will be satisfied with the phenomenological relation (2.1). In order to penetrate more deeply into the mechanism of spike generation, the Hodgkin–Huxley equations are used (see Sect. 11.1). These equations describe the generation of action potentials caused by the in- and outflux of ions. Depending on the kind of ions and their channels, extensions of these equations have also been developed and we will briefly represent them in Chap. 11. There we will also discuss the FitzHugh–Nagumo equations that allow us to get some insight into the nonlinear dynamics that produces spikes. From a physicist's point of view, neurons are by no means passive systems in thermal equilibrium. Rather they may be compared to machines that perform specific tasks, for instance the conversion of a constant signal into spikes, whereby the spike rate encodes information. When speaking of machines, we usually think of highly reliable performance; this is not the case with neurons, however. Due to fundamental physical principles, we must expect them to be rather noisy. We will study the generation of noise, both in dendrites and axons, in Sect. 4.8.

So far, in this chapter, we have been dealing with a single neuron. In Chap. 3 we will discuss some important aspects of their cooperation.

2.6 Learning and Memory

Though in our book we will not directly be concerned with processes of learning, a few comments may be in order, because they are linked with the existence of neurons. According to a widely accepted hypothesis due to D. O. Hebb, learning rests on a strengthening of the synapses that connect those neurons that are again and again simultaneously active, and similarly on a decrease of synaptic strengths if one or both neurons are inactive at the same time. In particular, Eric Kandel studied and elucidated the connection between the learning of behavioral patterns and changes at the neural level, in particular in sea slugs, such as Aplysia and Hermissenda.

Let us finally discuss the role of dendrites.

2.7 The Role of Dendrites

Dendrites are thin fibers along which ions may diffuse, thus generating an electric current. Such diffusion processes in one dimension are described by the cable equation. While it was originally assumed that the signal is transmitted from a synapse to the soma, more recent results show that back flows may also occur. Diffusion is a linear process. More recent theoretical approaches also consider nonlinear effects similar to the propagation of axonal pulses. Because of the transport of electric charges in dendrites, they give rise to electric and magnetic fields. Such fields stemming from groups of neurons can be measured using EEG (electroencephalograms) and MEG (magnetoencephalograms).

3. Neuronal Cooperativity

3.1 Structural Organization

The local arrangements of neurons and their connections are important for their cooperation. Probably the best studied neuronal system in the brain is the visual system. Since a number of important experiments that concern the cooperation of neurons have been performed on this system, we will briefly describe it in this section. At the same time, we will see how this organization processes visual information. So let us follow up the individual steps.

We will focus our main attention on the *human* visual system, but important experiments have been performed also on cats, monkeys and other mammals as well as on further animals, which we will not consider here, however. Light impinging on an eye is focussed by means of its lens on the retina. The latter contains rods and cones, whereby the rods are responsible for black and white vision, while the cones serve colour vision. In order to bring out the essentials, we present basic results on the rods. At their top, they contain membranes, which, in turn, contain a specific molecule called rhodopsin, which is composed of two parts. When light hits the molecule it decays, whereby a whole sequence of processes starts that, eventually, changes the permeability of the outer membrane of the rod. In this way, the potential between the inner and outer sides of the rod changes. Actually, even in darkness, i.e. when the rod is at rest, there is already a potential present. The inner side is slightly positively charged as compared to the outer side. The voltage is about 30–40 mV. When light impinges on the rod, its voltage is increased. Actually, this is in contrast to what is found in other sensory cells, where this potential is diminished. The more intense the impinging light, the stronger this voltage change, that continues until no more light comes in. The intensity of the impinging light is translated into an electrical excitation. This transformation requires energy that is delivered by chemical energy stored in the cells. By means of that energy, the degraded rhodopsin can be regenerated and is again available. Besides rods and cones, the retina contains further types of cells. We will not deal here with them in detail; may it suffice to mention that these cells interact with their neighbours both in the lateral as well the vertical direction of the retina, whereby information is carried on by means of voltage changes. The outer layer of the retina contains the *ganglion* cells that convert voltage changes into pulses. A certain array of

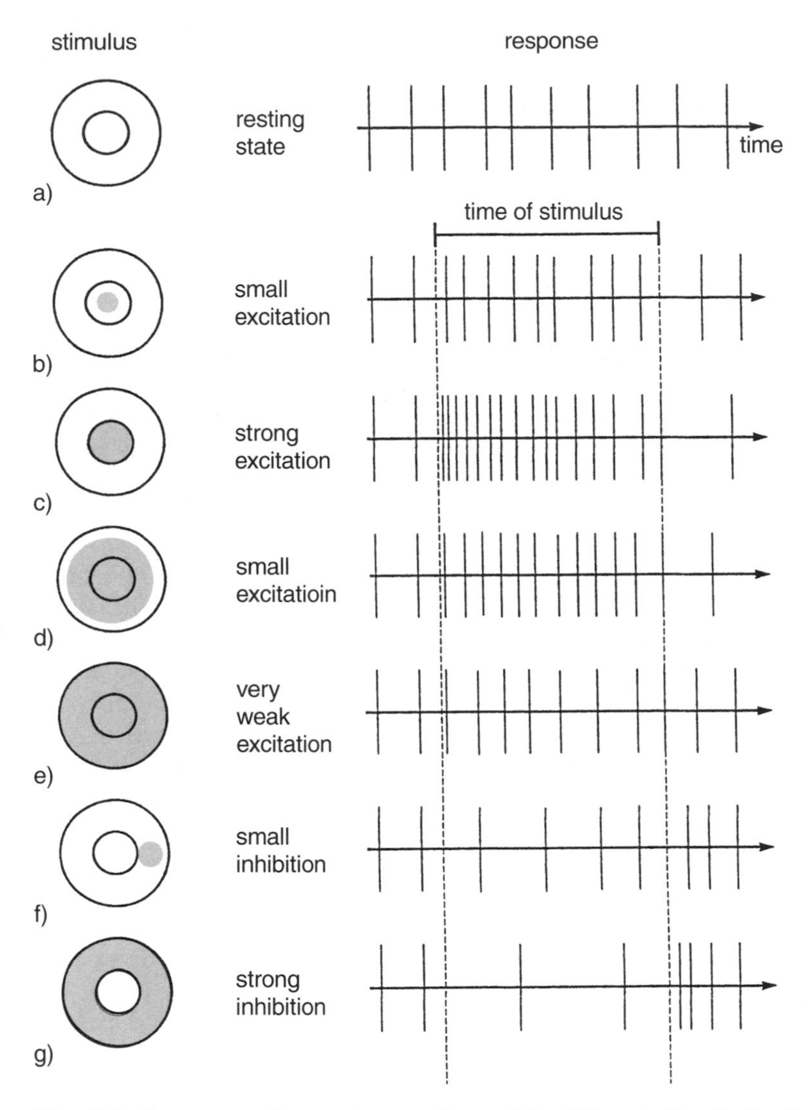

Fig. 3.1. Responses of on-center ganglion cell to different stimuli. On the l.h.s. the receptive field that consists of the center and surround. The light stimulus is represented dark. The r.h.s. represents the response of the neuron to the corresponding stimulus

rods contributes to the activity of a specific ganglion cell. The corresponding area covered by the rods is called the *receptive field* of the ganglion cell. The receptive fields of these cells are of a circular shape. *The receptive field* is a central concept that will accompany us through the whole visual system. Readers interested in the details of excitation and inhibition of ganglions are referred to Fig. 3.1 and its legend. While in the so-called on-cells the center of the receptive field leads to an excitation of the cell and an inhibition at its periphery, in off-cells just the opposite occurs.

The nerve pulses are conducted along nerve fibers to a change point, where some of the nerve fibers change from one side of the brain to the other (Fig. 3.2). Then they go on to the corpus geniculatum laterale. There some kind of switching occurs and the nerve fibers further proceed to the visual cortex at the rear part of the brain. Other fibers go to other parts of the brain. Behind the change point one half of the ongoing nerve consists of nerve fibers that stem from the right eye and the other half of nerve fibers from the left eye. Nerve fibers that stem from the left parts of the retinas of both eyes go on to the left brain, whereas nerve fibers that stem from the right parts of the retinas go on to the right half of the brain. When we take into account that the image that is perceived by humans on their retinas is mapped upside-down and the sides interchanged, it follows that on the right halves of the retinas the left part of the visual field is perceived and vice versa. For instance, when on the left-hand side of a table there is a ball and on its

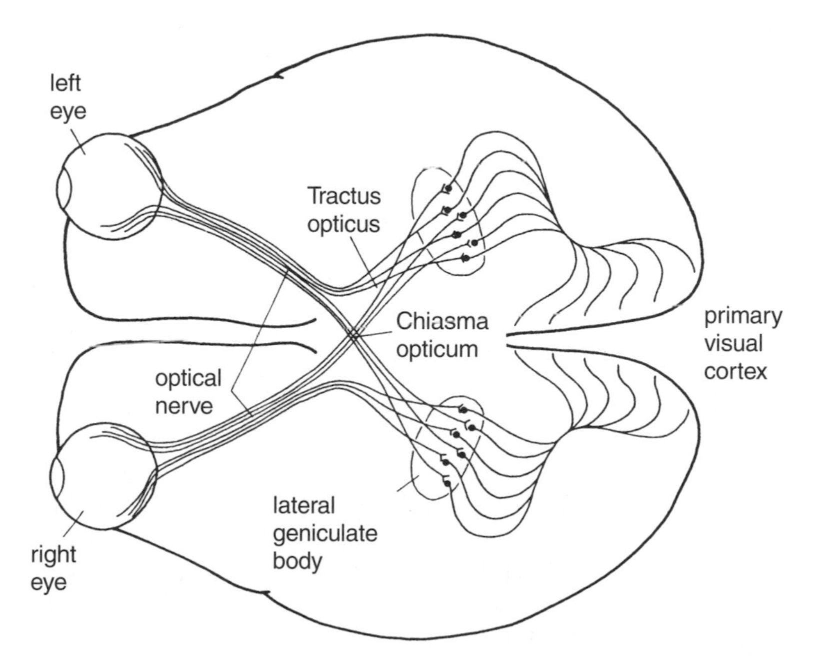

Fig. 3.2. Schematic representation of the visual pathway of a human

right-hand side a pencil, the pencil will be mapped onto both left retinas and the information is transferred into the left part of the brain. The ball lying on the right side is perceived by the right part of the brain. Both parts of the visual field are thus processed separately in both halves of the brain. On the other hand, each half of the brain receives information from both eyes. This actually serves stereovision, because in this way each half of the brain is able to process the images that stem from both eyes jointly.

It should be noted that the spatial order is conserved in the whole visual system; that is to say that nerve fibers that deliver information from neighbouring regions of their retina always remain neighbours. The local scales of these topological maps are not conserved, however. For instance, the map of the "yellow spot" of the retina possesses a comparatively large area in the corresponding part of the brain. The nerve fibers proceed from the corpus geniculatum laterale (Fig. 3.2) to the *primary visual cortex* from where connections exist to a number of layers. Numerous connections exist to other brain areas, for instance to a reading center from where information can be passed on to a speech center, and so on.

In the primary visual cortex a white stripe in the otherwise gray cortical substance is particularly easily visible. The stripe is named the Gennari stripe after its discoverer and the corresponding brain area is called the *striped cortex* or *area striata*. The visual cortex is a sheet of cells about 2 mm thick and with a surface of a few square centimeters. It contains about 200×10^6 cells. Neurologists distinguish between different subunits of the area of the cortex that processes visual perception. The first station, where the fibers of the lateral geniculate body terminate, is the so-called primary visual field. This is also called *area 17* or *V1*. This area is followed by areas that are called 18, 19, etc. or *V2, V3*, etc. For our purposes it will be sufficient to deal with a rough subdivision into a primary visual field and secondary or higher visual fields. It is important to note, however, that each visual field represents a more or less complete representation of the retina. In other words, excitation of a certain area of the retina causes a response in a definite area of this visual field. Thus, the visual field represents a map of the retina. Of course, we must be aware that in each half of the brain each time only half of the retinas of both eyes are mapped. Today it is estimated that the cortex of monkeys contains at least 15 different visual fields and possibly in humans there are still more. Only the primary visual field has been well studied up to now and we will present some of the results here. The cortex in the region of the primary visual field can be subdivided into six layers that differ with respect to the types of cells they contain and also with respect to the density of cells. These layers are numbered from I to VI with further subdivisions. Nearly all nerve fibers from the lateral geniculate body terminate in layer IVc. It must be noted that information is not only processed in one direction, but there are also a number of back propagations.

Let us discuss in which way neurons in the visual cortex react to receptive fields. Actually, there are quite a number of different cells that react to different excitations. Neurologists differentiate between simple and complex cells. As these notations indicate, the receptive fields of different cells differ with respect to their complexity. The receptive fields of nerve cells in the visual cortex were mainly studied in cats and monkeys. Remember that in each case the receptive fields refer to the retina, i.e. a small region of the retina influences the corresponding neuron in the visual cortex. In it there are cells that possess circular receptive fields with a center and an oppositely acting surround. These cells are located exclusively in area IVc and are all monocular, i.e. they are fed only from one eye. It is assumed that these cells represent the first station in the visual cortex. But most of the so-called simple cells don't possess circular receptive fields, they are actually rectangular (see Fig. 3.3). Basically, these cells are very sensitive to the direction of a bar. Only bars with a specific direction cause an optimal response in the cell. There are neurons for each direction, whereby, for instance, neither vertical nor horizontal is preferred. It is assumed that the properties of these cells are brought about by the cooperation of simpler cells with circular receptive fields. The simple cells have in common that they possess well-defined excitatory and well-defined inhibitory fields. In all cases excitations that don't change in time suffice to excite the simple cells.

The situation is quite different in complex cells. The receptive fields of complex cells are larger that those of the simple cells and they can't be divided into clearly defined excitatory and inhibitory zones. The *complex cells* are characterized by the fact that they react in particular to *moving excitations*, especially to light bars that move perpendicularly to their extension. Complex cells exhibit a specific orientation sensitivity, i.e. only a correctly oriented bar that is moving in the corresponding direction leads to a response in the corresponding cell.

So far we have got acquainted with the most important cells of the primary visual field. These cells are arranged in strict organization by means of columns. Each column is about 30–100 μm thick and 2 mm high. Each of these columns contains cells of the fourth layer with circular receptive fields. Above and below each, simple and also complex cells can be found. What is particularly interesting is the fact that all orientation-specific cells of a column react to the same orientation of a light bar (Fig. 3.4). Neighbouring columns differ with respect to their orientation specificities by about 10°. Thus going from one side to the other of a column, we find a slight change of orientation from initially vertical to finally horizontal orientation. We may distinguish between columns that are mainly served from the left or from the right eye so that they are called *ocular dominance columns*. Each small section of the retina thus possesses a corresponding set of columns with all possible directions and for both eyes. According to Hubel and Wiesel such a set is called a *hyper-column*. Nearly similar to a crystal, in the visual cortex

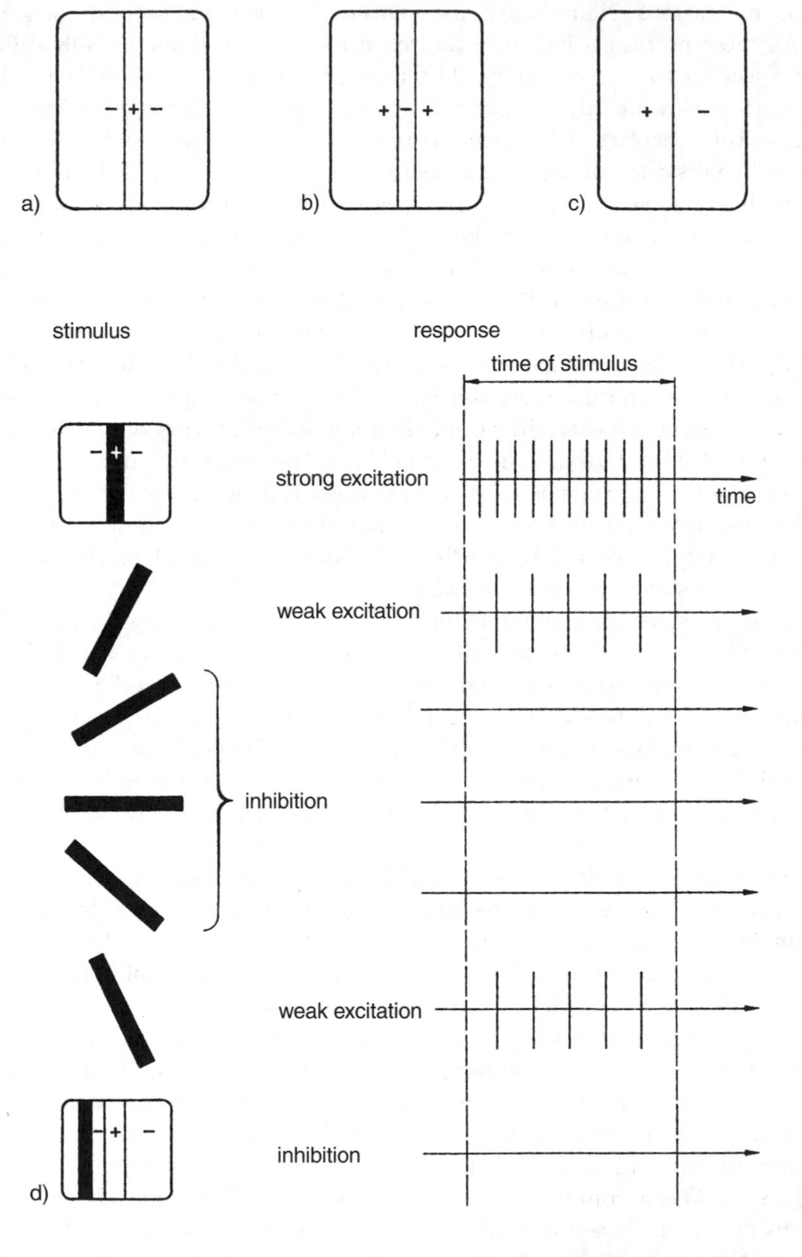

Fig. 3.3. Response of a neuron with a specific receptive field to a light bar with different orientations. The receptive field is shown in the *upper left square*. + marks the region of excitation, – that of inhibitation in the case of illumination

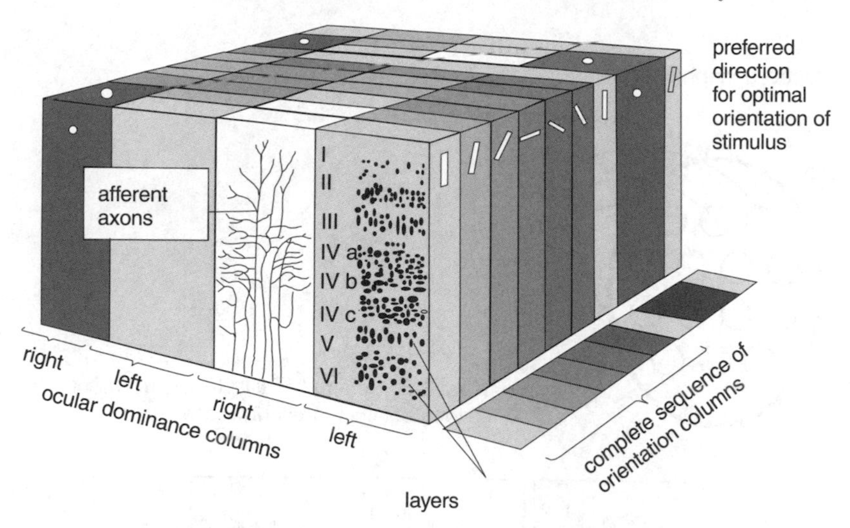

Fig. 3.4. Organisation of visual cortex in the form of columns

such hyper-columns are regularly arranged and can be attributed to a small area of the retina. A similar orientation in the form of columns can also be found in other parts of the cortex. Cells in still higher visual fields have still more complex properties; that is why they are called *hyper-complex*. Earlier hypotheses on brain functioning assumed that, eventually, there are specific cells that recognize specific objects and jokingly such cells were called *grandmother cells*. Now the conviction has won that such grandmother cells don't exist and that the recognition, say of a specific face, is the result of the cooperation of many individual cells. It is here where the question of neuronal cooperativity becomes especially important and we will discuss experiments and some hypotheses later in this chapter.

3.2 Global Functional Studies. Location of Activity Centers

There are a number of methods that allow us to study experimentally in which regions of the brain specific mental tasks are performed. These methods rest on the assumption that during increased mental activity, the blood flow in the active region is increased. Clearly, the time constants are rather large and lie in regions above seconds. This is a region that we are not interested in in our later models, so we mention the corresponding methods only briefly. These are positron emission tomography (PET), magnetic resonance imaging (MRI), better known among physicists as nuclear magnetic resonance (NMR), and, with a higher temporal resolution, functional magnetic resonance imaging (fMRI). An interesting method that was introduced more recently by Grin-

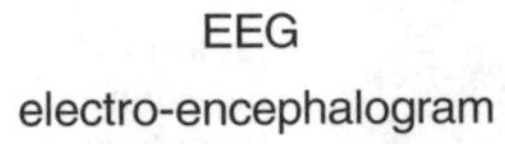

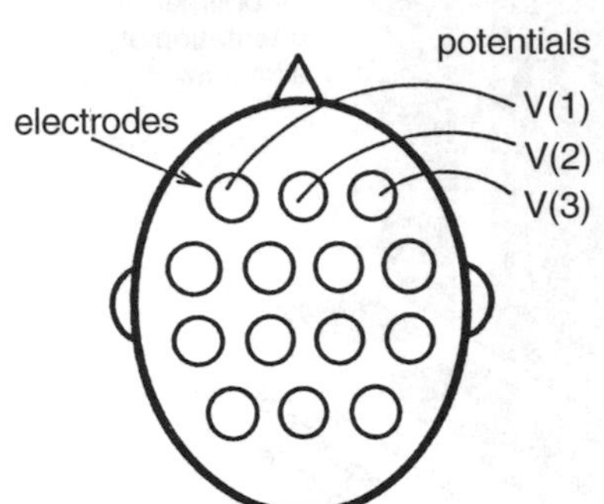

Fig. 3.5. Scheme of EEG measurement. Positions of electrodes on scalp

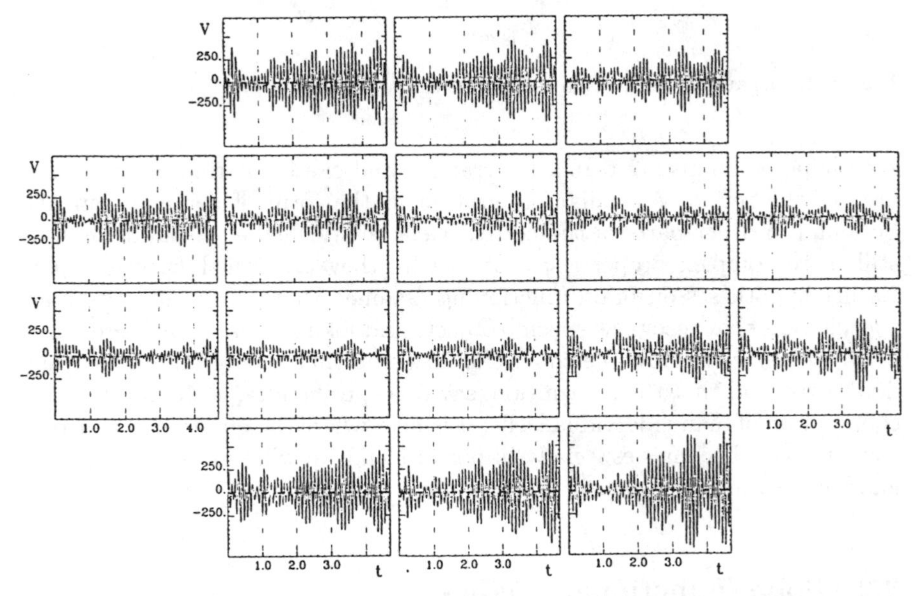

Fig. 3.6. Scheme of representation of EEG measurement. Each box is a plot of voltage V versus time, corresponding to the electrodes of Fig. 3.5 (after Lehmann, private communication)

vald is optical imaging of the visual cortex that requires, however, the opening of the skull of the animal. In view of the models we will discuss in our book, we first mention other methods that allow us to study rhythmic or oscillatory phenomena of large groups of neurons. These are the electroencephalogram (EEG) and the magnetoencephalogram (MEG). In the electroencephalogram electrodes are placed on the scalp and the voltages between these electrodes and a reference electrode are measured and plotted in ways shown in Figs. 3.5 and 3.6. Such EEGs may be measured either with respect to specific parts of the brain, e.g. the sensory or the motor cortex, or with respect to the total scalp. In general, specific frequency bands are filtered out and studied, for instance the α-band in the region from about 8–12 Hz, or the γ-band, and so on. In general, these bands are associated with specific global activities of the brain, for instance sleep, attention, and so on. By means of squids (superconducting quantum interference devices) it is possible to measure very small magnetic fields.

3.3 Interlude: A Minicourse on Correlations

One of the most exciting developments in experimental brain research is the discovery of pronounced *temporal correlations* between the activity of neurons. Here it has been important that the experimental results can be represented in a rigorous mathematical form. Thus, before we describe the experimental results, we will have to discuss a number of basic concepts, whereby we focus our attention on spikes. The occurrence of a spike is, in a technical sense, an *event*. In the following, we will be concerned with the *statistics* of such events. We may record events either by measuring many neurons simultaneously or the same neuron in repeated experiments. Because the experimental time resolution is finite, we consider discrete time intervals Δ and define (Fig. 3.7)

$$t_n = n\Delta - \Delta/2, \quad n = 1, 2, ... \tag{3.1}$$

Experimentally, we may count the *number of events* in the interval Δ at time t_n (Fig. 3.8a) and call that number (Fig. 3.8b)

$$N(t_n) \equiv N_n. \tag{3.2}$$

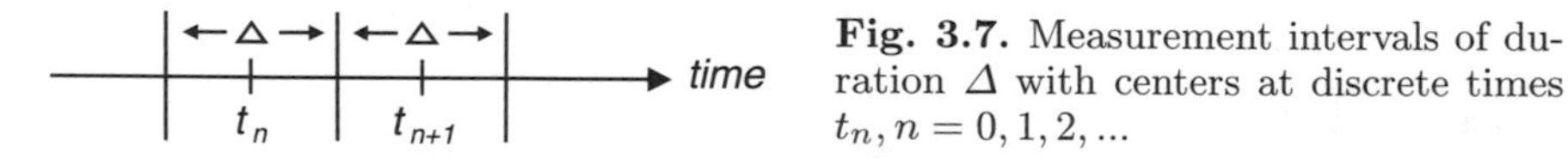

Fig. 3.7. Measurement intervals of duration Δ with centers at discrete times $t_n, n = 0, 1, 2, ...$

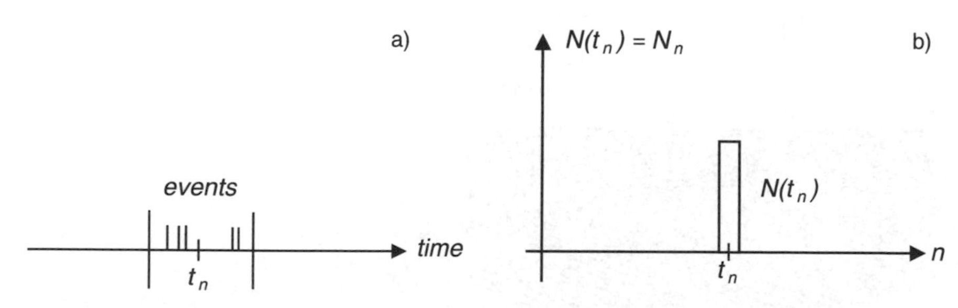

Fig. 3.8. **a**) Events within a measurement interval with center at t_n; **b**) Number $N(t_n) = N_n$ of events within measurement interval with center at t_n

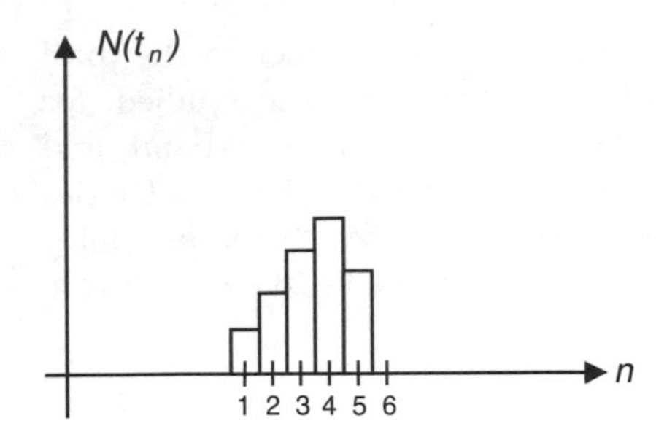

Fig. 3.9. Example of a histogram

The plot of (3.2) versus n is called a *histogram* (Fig. 3.9).
The number N_n is also called the *frequency of events*. When we divide it by the total number of events

$$N = \sum_n N_n \,, \tag{3.3}$$

we obtain the *relative frequency*

$$p_n = N_n/N \,. \tag{3.4}$$

We now remind the reader of the concept of the *average* of a quantity $x(t)$. For example, $x(t)$ may be the postsynaptic potential $x_j(t)$ measured at time t at neuron j, $j = 1, 2, ..., J$. The ensemble average is defined by

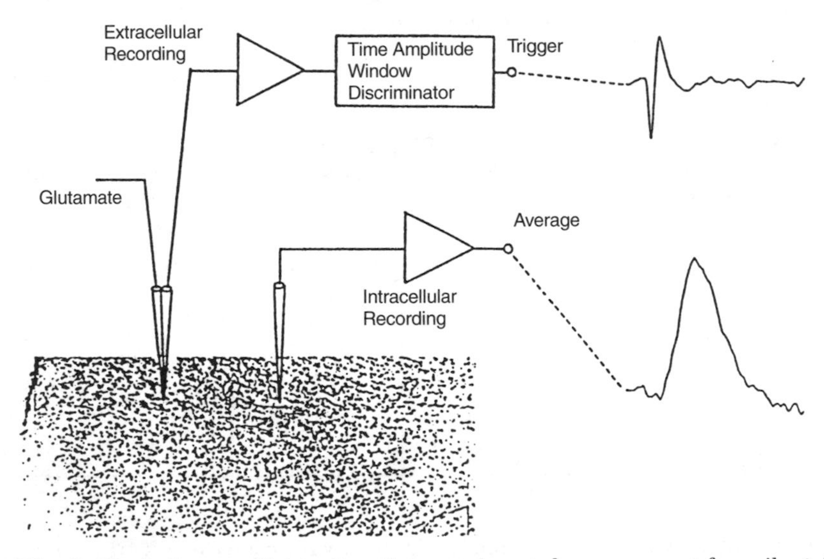

Fig. 3.10. A diagram illustrating the experimental arrangement for spike-triggered averaging in slice preparations of the visual cortex (after K. Toyama, in Krüger, 1991)

$$< x(t) >= \frac{1}{J} \sum_{j=1}^{J} x_j(t) . \tag{3.5}$$

This definition, which may either be applied to many neurons or to repeated experiments on the same neuron, leads to a difficulty in practical applications in neural science, because the individual curves $x_j(t)$ may be time-shifted with respect to each other so that the average is more or less wiped out, whereas the curves are of quite similar shape but differ just by time-shifts. Thus we need a marker, which can be given by a neuronal spike, so that the beginnings of the individual tracks refer to the same initial signal (Figs. 3.10 and 3.11). Such averages are called *spike-triggered averages*

$$< x(t) >= \frac{1}{J} \sum_{j=1}^{J} x_j(t - \tau_{trigger}) . \tag{3.6}$$

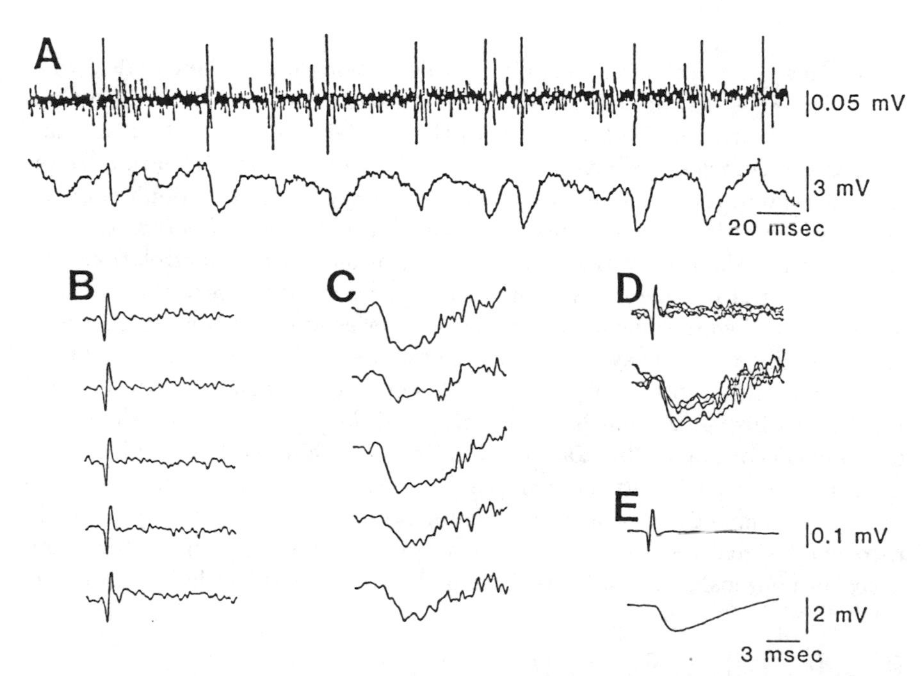

Fig. 3.11. Simultaneous recording from two visual cortical cells and spike triggered averaging.
A) Extracellular recording (*upper trace*) from juxta-granular (JG) layer; intracellular recording of postsynaptic potentials from the other cell in the supragranular layer (SG).
B) Extracellular impulses in a JG cell isolated by a time-amplitude window discriminator.
C) Intracellular traces in a SG cell triggered by the JG impulses.
D) and E) Superimposed and averaged traces in JG and SG cells (after Komatsu et al., 1988; reproduced by Toyoma, in Krüger, 1991)

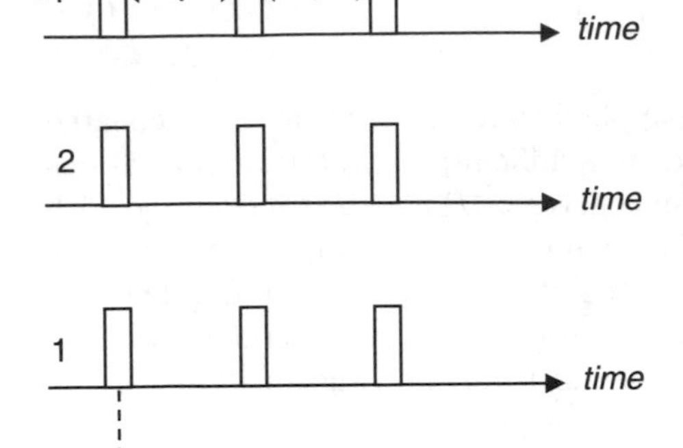

Fig. 3.12. Two synchronous spike trains with equal spike intervals T (schematic)

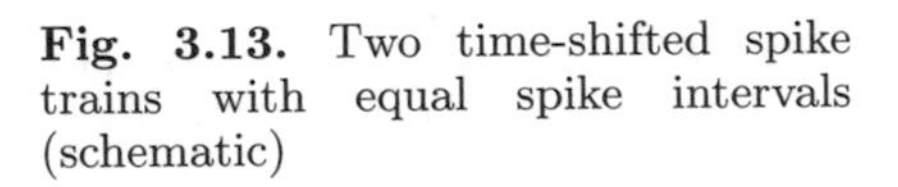

Fig. 3.13. Two time-shifted spike trains with equal spike intervals (schematic)

We now come to the central point of this section, namely how to deal with *correlations*. Let us consider the example of two neurons 1 and 2, which both produce spike trains (see Fig. 3.12). In the special case where the trains are regular and coincide, we have perfect *synchrony* or perfect correlation. If, on the other hand, the two spike trains don't perfectly coincide or don't coincide at all, the correlation is diminished. Clearly, the number of *coincidences* of spikes during the observation time t_{obs} is a measure of the correlations. As shown in Fig. 3.13, there may be no coincidences at all between the two time series, but there are coincidences if the time series are shifted with respect to each other by a time delay τ. Thus it suggests itself to consider the number of coincidences as a function of τ and plot this number N_τ^c against τ. In this way, we obtain a histogram that is also called a *correlogram*. Such a correlogram is evaluated according to the example of finite spike trains as shown in Fig. 3.14.

Let us study, as a further example, the correlogram that refers to a spike train and a signal that consists of many uncorrelated spikes so that practically a constant signal results. In the present context, this signal may stem from a *pool* of neurons. If the spike train and the constant signal last over a time

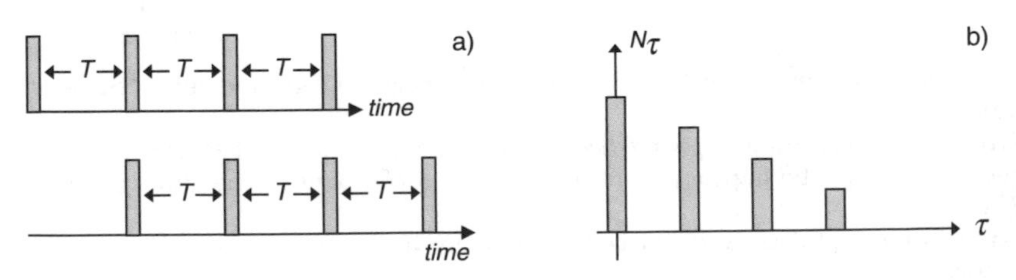

Fig. 3.14. a) Two time-shifted spike trains with equidistant spikes and finite length (schematic); **b**) Correlogram corresponding to a)

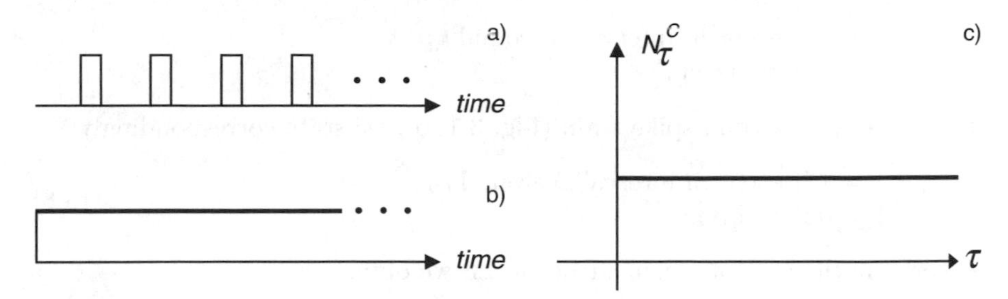

Fig. 3.15. **a**) infinite spike train with equidistant spikes (schematic); **b**) constant signal; **c**) Cross-correlogram corresponding to the above signals

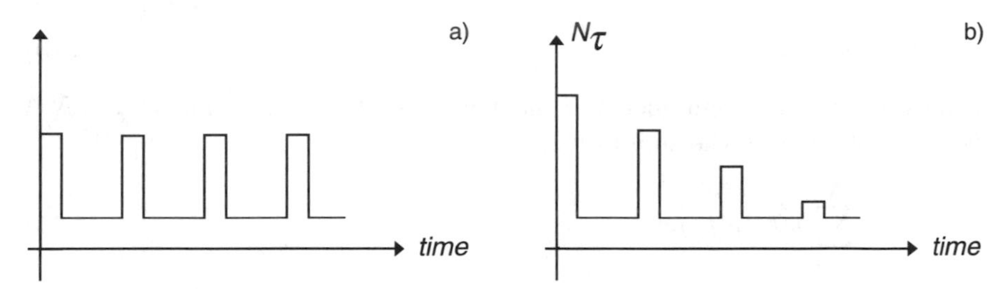

Fig. 3.16. **a**) Spike train of finite length with constant underground signal (mimicking noise) (schematic); **b**) (Auto-) Correlogram corresponding to a)

that is considerably larger than the delay τ, the number of coincidences is practically constant and the correlogram looks like that of Fig. 3.15c. An *autocorrelogram* is obtained if we correlate two timeshifted signals that stem from the *same pool.* If the signal is composed of regular spikes and the same background (Fig. 3.16a), the autocorrelogram of Fig. 3.16b results. Further important concepts are those of the *autocorrelation function* and the *cross-correlation function.* To pave a way to these concepts, we first show how we can calculate the number of coincidences between two spike trains in a more formal fashion. We divide the total observation time t_{obs} into intervals Δ so small that only *one* event (spike) or no spike falls into such an interval (Fig. 3.17a). We then state

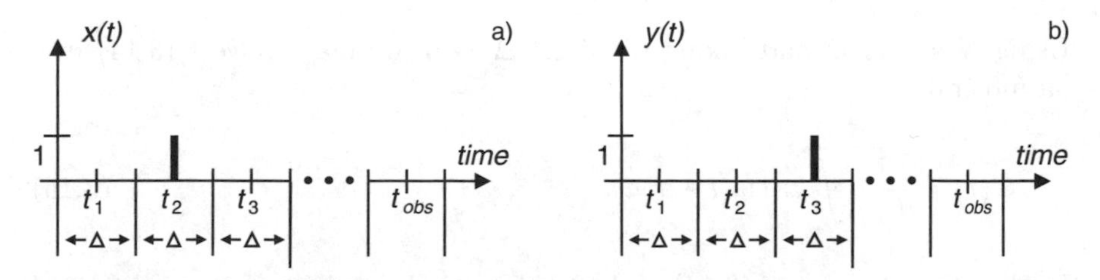

Fig. 3.17. Correlation of two spikes with heights $x(t)$ and $y(t)$, respectively. For details see text

$$x(t_n) \begin{array}{l} = 1 \text{ if event in interval } \Delta \text{ around } t_n\,, \\ = 0 \text{ if no event}\,. \end{array} \tag{3.7}$$

If we observe a second spike train (Fig. 3.17b), we state correspondingly

$$y(t_n) \begin{array}{l} = 1 \text{ if event in interval } \Delta \text{ around } t_n\,, \\ = 0 \text{ if no event}\,. \end{array} \tag{3.8}$$

Clearly, in the case of a coincidence in Δ, we obtain

$$x(t_n)y(t_n) = 1\,, \tag{3.9}$$

and

$$x(t_n)y(t_n) = 0 \tag{3.10}$$

otherwise. The total number of coincidences in observation time $t_{obs} = N\Delta$ is then obtained by the sum over t_n, i.e.

$$N^c = \sum_{n=1}^{N} x(t_n)y(t_n)\,. \tag{3.11}$$

If $y(t)$ is time-shifted by an amount τ, the total number of coincidences in observation time t_{obs} becomes

$$N_\tau^c = \sum_{n=1}^{N} x(t_n)y(t_n+\tau)\,, \tag{3.12}$$

where

$$y(t_n+\tau) = 0 \quad \text{if } t_n+\tau \text{ outside observation interval}\,. \tag{3.13}$$

In (3.11) and (3.12) the variables $x(t_n), y(t_n)$ can adopt only the values zero or one. If we allow these variables to adopt arbitrary real values, and normalize (3.11) and (3.12) properly, we arrive at the definition of the *crosscorrelation function* for discrete intervals

$$C(\tau) = \frac{1}{N}\sum_{j=1}^{N} x(t_j)y(t_j+\tau)\,. \tag{3.14}$$

Using $N = t_{obs}/\Delta$ and taking the limit $\Delta \to 0$, we may convert (3.14) into an integral

$$C(\tau) = \frac{1}{t_{obs}} \int_0^{t_{obs}} x(t)y(t+\tau)dt\,. \tag{3.15}$$

In the *stationary* case, (3.14) and (3.15) are independent of the position of the time window. More generally, when the time window starts at time t and extends to $t + t_{obs}$ in the nonstationary state, we obtain

$$C(t,\tau) = \frac{1}{t_{obs}} \int\limits_{t}^{t+t_{obs}} x(\sigma)y(\sigma+\tau)d\sigma\,, \tag{3.16}$$

where C explicitly depends on t. In (3.14)–(3.16) the correlation function is defined by means of a time average, i.e. the sum or integral run over time. We may equally well define an ensemble average in analogy to (3.5) by means of

$$C(t,\tau) = \sum_j x_j(t)y_j(t+\tau)\,. \tag{3.17}$$

The auto-correlation function is obtained as a special case from (3.14)–(3.17) by putting $y(t) = x(t)$ or $y_j(t) = x_j(t)$.

3.4 Mesoscopic Neuronal Cooperativity

In this section we want to present some of the fundamental experimental results on the synchronization among groups of neurons. Such synchronizations were discovered by Walter Freeman in the olfactory bulb of rats, but particularly striking experiments were performed by Singer et al. as well as by Eckhorn et al. on the visual cortex of anaesthesized cats and later on awake monkeys. Let us quote the discoverers. Singer (1991): "We have discovered that a large fraction of neurons in the cat striate cortex engage in oscillatory activity in a frequency range of 40 to 60 Hz when activated with light stimuli to which the neurons are tuned (Gray and Singer, 1987; Gray and Singer, 1989). This phenomenon is illustrated in Fig. 3.18. Units close enough to be recorded with a single electrode, if responsive to the same stimulus, always synchronize their respective oscillatory responses. In most instances, oscillatory responses are also in phase for neurons aligned along the vertical axis of a cortical column. Of particular interest in the present context is the finding that the oscillatory responses can also synchronize over considerable distances across spatially separate columns (Gray et al., 1989) and even between cortical areas (Eckhorn et al., 1988). Thus far, three parameters have been identified which determine the degree of synchrony within area 17: the distance between the units, the similarity of their orientation preference, and the coherence of the stimulus itself. When neurons are less than 2 mm apart, in which case the receptive fields are usually overlapping, they always synchronize their oscillatory responses when they show the same orientation preference, and they often synchronize even if the orientation preferences differ, as long as these differences are sufficiently small to allow activation of both neuron clusters with a single stimulus. At larger distances, when the receptive fields are no longer overlapping, cell clusters tend to synchronize their oscillatory responses only when they have similar orientation preferences and/or are activated by stimuli that have the same orientation and that move in the same direction. In such cases correlation breaks down when the three

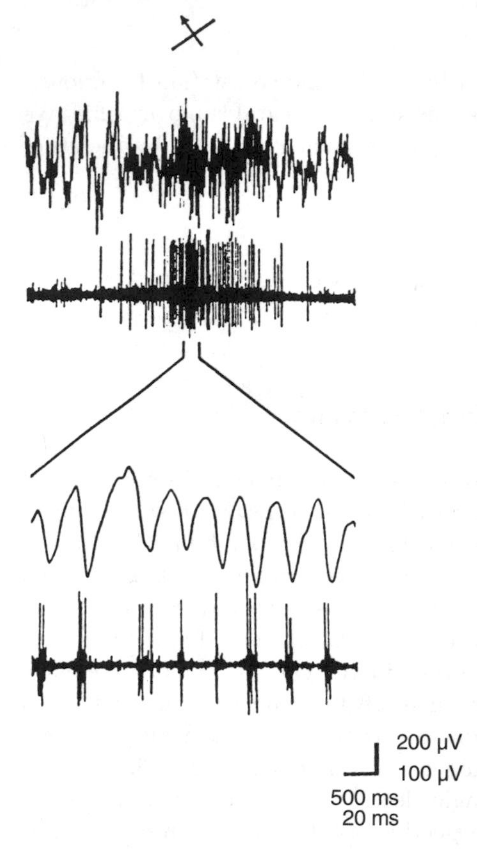

Fig. 3.18. Time series of multiunit activity (MUA) and local field potential (LFP) in the first and second row, respectively. The *lower two rows* show an enlarged section. The MUA and LFP responses were recorded from area 17 in an adult cat due to the presentation of an optimally oriented light bar moving across the receptive field. Oscilloscope records of a single trial showing the response of the preferred direction of movement. In the *upper two traces* at a slow time scale the onset of the neuronal response is associated with an increase in high-frequency activity in the LFP. The *lower two traces* display the activity of the peak of the response at an expanded timescale. Note the presence of rhythmic oscillations in the LFP and MUA (35–45 Hz) that are correlated in phase with the peak negativity of the LFP. Upper and lower voltage scales are for the LFP and MUA, respectively (from Gray and Singer, 1989)

stimuli pass in opposite directions over the two receptive fields and reaches its maximum when both neuron clusters are activated with a single continuous stimulus (Gray et al., 1989).

This phenomenon is illustrated in Fig. 3.19. Under each of the three stimulation conditions, the autocorrelations show a periodic modulation indicating that the local responses are oscillatory. However, when the two stimuli move in opposite directions, where they are perceived as two independent contours, the cross correlation function is flat implying that the respective oscillatory responses have no consistent phase relation. When the two stimuli move in the same direction, where they are perceived as related, the cross correlogram shows a periodic modulation indicating that the respective oscillatory responses are in phase. Synchronization improves further when the two stimuli are replaced by a single bar of light. This can be inferred from the deep modulation of the oscillatory cross correlogram in Fig. 3.19. Interestingly, if cross correlograms are periodically modulated, they are always centered

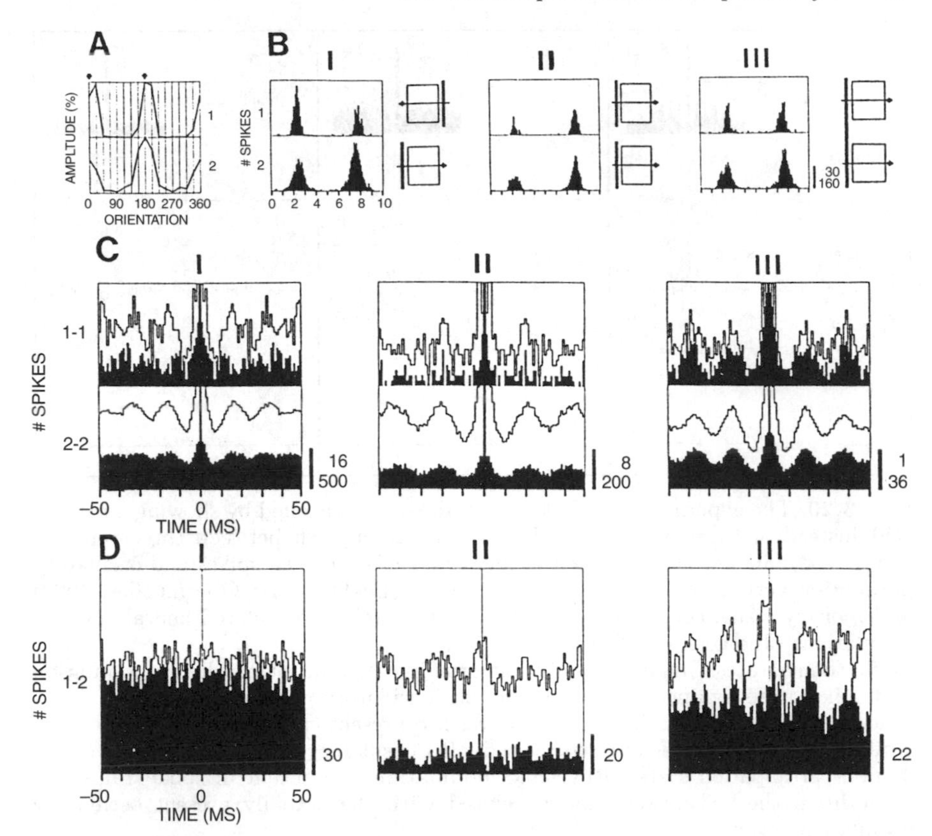

Fig. 3.19. Long range of oscillatory correlations reflect global stimulus properties. **A:** Orientation tuning curves of neuronal responses recorded by two electrodes (1,2) separated by 7 mm show a preference for vertical light bars (0 and 180^0) at both recording sides.
B: Post-stimulus time histograms of the neuronal responses recorded at each site for each of three different stimulus conditions: (I) two light bars moved in opposite directions, (II) two light bars moved in the same direction and (III) one long light bar moved across both receptive fields. A schematic diagram of the receptive field locations and the stimulus configuration used is displayed to the right of each post-stimulus time histogram. **C, D:** autocorrelograms (C, 1-1, 2-2) and cross-correlograms (D, 1-2) computed for the neuronal responses at both sides (1 and 2 in A and B) for each of the three stimulus conditions (I, II, III) displayed in B. For each pair of correlograms, except the two displayed in C (I, 1-1) and D (I), the second direction of stimulus movement is shown with unfilled bars. Numbers on the vertical calibration correspond to the number of coincident events (spikes) in the respective auto- and crosscorrelograms (from Gray et al., 1989)

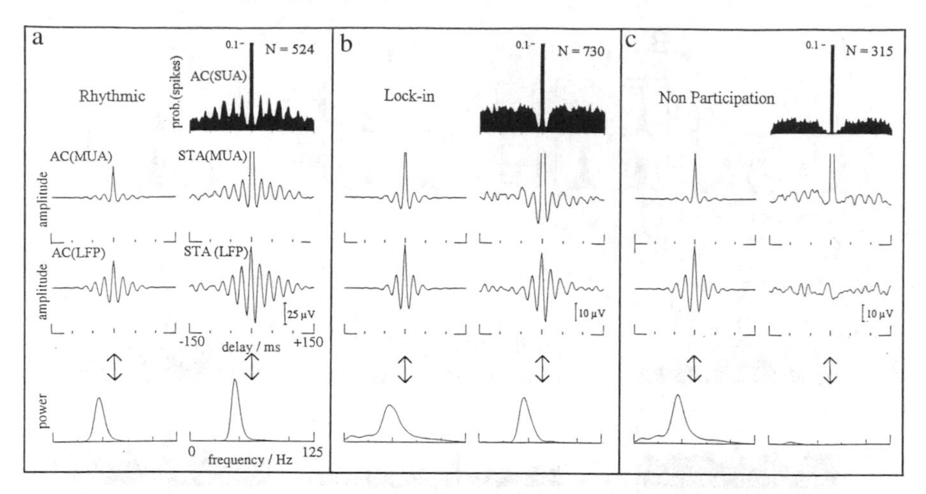

Fig. 3.20. The experiments by Eckhorn et al. were performed by showing a moving grid instead of the moving bar. The authors distinguish between three types of coupling dynamics, based on their observations of single cell spikes and oscillatory population activity of LFP and MUA in cats (Eckhorn and Obermueller, 1993) and monkey visual cortex (Frien et al., 1994) as well as on related neural network simulations (Juergens and Eckhorn, 1997).
1) rhythmic states. Single cell spike patterns had significant rhythmic modulation and spikes were significantly correlated with oscillatory population activity.
2) lock-in states. Rhythmic modulation was not present in single cell spike patterns, while spikes were significantly correlated with oscillatory population activity.
3) non-participation states. Rhythmic modulation was absent in spike trains and in addition the spikes were not correlated with the actually present oscillatory population activity.
This figure shows the results for the three different states of single cell couplings with oscillatory population activities in the primary visual cortex:
a) rhythmic;
b) lock-in;
c) non-participation states of three different neurons.
AC: autocorrelation histograms (correlograms) of single cell spikes (SUA), of multiple unit activity (MUA) and of local field potential (LFP). STA denotes spike-triggered averages of multiple unit activity or local field potentials. According to the classification, STAs have oscillatory modulations in the rhythmic and lock-in states and lack such modulations in the non-participation state. Note that in the rhythmic state (a) the single cell correlogram (*top*) is clearly modulated at 44 Hz, while in the lock-in (b) and the non-participation states (c) rhythmic modulations in the range 35–80 Hz are not visible (by definition).
Lower row of panels: power spectra for the above row of correlograms (figure modified from Eckhorn and Obermueller, 1993)

around zero phase angle. This indicates that oscillations tend to synchronize in phase if they synchronize at all".

Eckhorn (2000) describes his experiments as follows: "Our experimental evidence is based on multiple microelectrode recordings from the visual cortex of anaesthesized cats and awake moneys. We have used local population activities (multiple unit spike activity (MUA)) and local slow-wave field potentials (1:150 Hz) (LFP)) for the analysis of cortical synchrony, because they comprise the synchronized components of local populations (e.g. Eckhorn, 1992). In particular, LFPs are a local weighted average of the dendro-somatic postsynaptic signals, reflecting mainly the synchronized components at the inputs of the population within approximately 0.5 mm of the electrode tip (Mitzdorf, 1987). MUAs on the other hand, comprise, in their amplitudes, the simultaneity of spikes occurring at the outputs of a local population within approximately 0.05 mm of an electrode tip (Legatt et al., 1980, Gray et al., 1995). Both population signals are more suitable for detecting correlations among dispersed cortically recording locations than are single unit spike trains (Eckhorn et al., 1988, 1990, Eckhorn, 1992, Gray et al., 1989, 1990; Engel et al., 1990, 1991). Higher numbers of neurons contribute to LFP and to MUA due to their shallow spatial decay (Legatt et al., 1980; Mitzdorf, 1987)".

Clearly, both from the experimental and theoretical point of view, it will be ideal to measure the correlation between spike trains of two single cells. Under present experimental conditions the probability of finding such significantly coupled spike trains is, however, small. However, according to Eckhorn, the easier finding of cooperating neural assemblies was dramatically increased by recording neural mass signals, such as MUA and LFP in parallel with single unit spike trains. This *triple recording* from each electrode opened up for the researchers the opportunity to investigate neural couplings on different levels of organization and specificity, and analyze all possible combinations of interactions between the different signal types. By these means, the researchers can study the interactions between single neurons, possessing well-defined tuning properties, the average mass activity of the local dendritic slow-wave potentials (LFPs), and the spike activity of the local assembly (multiple unit activity MUA). Thus, correlations between the mass activities of the same or different types were evaluated. Using these different signal combinations, the researchers generally found significant and often strong signal correlations. Figures 3.18, 3.20 and 3.21 present typical results of Gray et al. and Singer et al. on the one hand, and Eckhorn et al. on the other.

After these fundamental discoveries, many further experiments were performed by Eckhorn and Singer and their groups, as well as by other researchers, but for our purpose it is sufficient to have characterized the basic phenomenon.

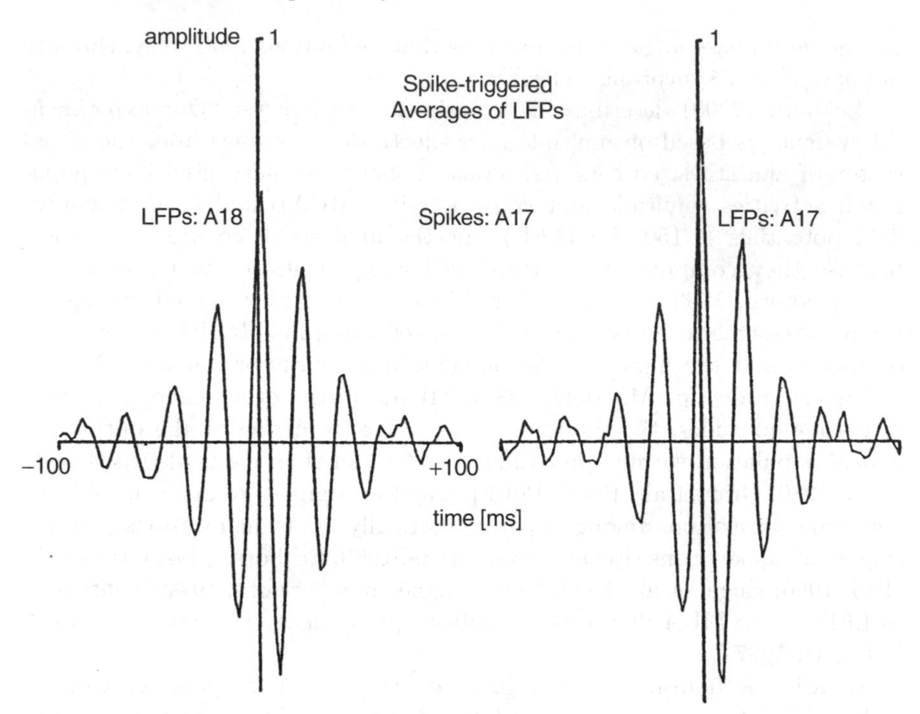

Fig. 3.21. Correlation between LFP oscillations and single cell spikes (SUA), calculated as spike-triggered average (STA) of LFPs normalized to the number of spikes and the energy of the respective LFPs. Negative LFP values plotted upwards; average oscillation frequency about 45 Hz. Spikes and LFPs were recorded in area 17 with the same electrode (right STA), and LFPs with another electrode in area 18 (left STA); i.e. the right STA shows the correlation between a single area 17 cell spike train and the LFPs from the direct neighborhood, while the left STA shows the correlation between the same area 17 cell and LFPs from area 18. (A 17 cell and A 18 receptive fields overlapped; stimulus: grating 0.7 cycles/° drifting at 8 °/s in and against preferred direction of cells in the area 17/area 18 recording positions.) (From experiments of R. Eckhorn, M. Munks, W. Kruse and M. Brosch quoted in Eckhorn, 1991)

Before I present several models of neural nets in which synchronization between the individual neurons may happen, I will outline important theoretical concepts and tools that will be needed in Part III in particular.

4. Spikes, Phases, Noise: How to Describe Them Mathematically? We Learn a Few Tricks and Some Important Concepts

In this chapter we present the ingredients that we will need to formulate our model on pulse-coupled neural networks. There are two ways of reading this chapter. The speedy reader will read Sect. 4.1 that shows how to mathematically describe spikes (or short pulses). Section 4.4 deals with a simple model of how the conversion of axonal spikes into dendritic currents at a synapse can be modeled. Finally, we will need the fundamental concept of phase that will be presented in Sects. 4.9.1 and 4.9.2. Combined with a knowledge of Chaps. 3 and 3 the reader will then easily understand the *lighthouse model* of Chap. 5 and its extensions. Later he or she may like to return to read the other sections of Chap. 4.

I personally believe that there is another class of readers who appreciate the systematic exposition and they will read all sections of Chap. 4. They will find details on how to solve the model equations (Sects. 4.2 and 4.3), how to model important phenomena of noise (Sects. 4.4–4.8), and how to extract phases from experimental data (Sect. 4.9).

4.1 The δ-Function and Its Properties

In this section we want to find a way to describe short pulses or, to use a different word, spikes. To this end, we will use an idealization that leads us to the δ-function originally introduced by Dirac. Let us first consider a pulse that has the form of a Gaussian bell-shaped curve (see Fig. 4.1). It is described

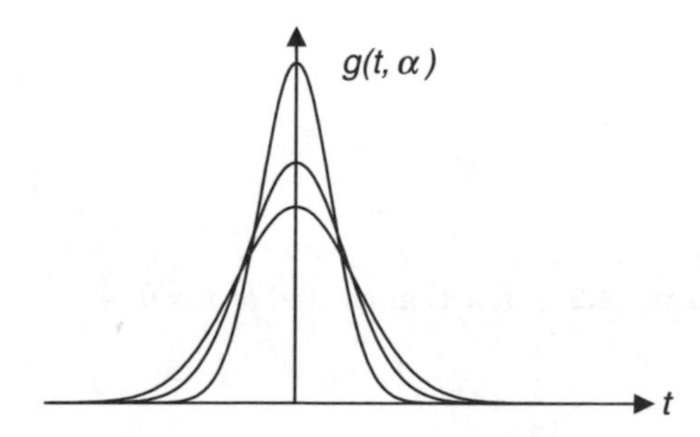

Fig. 4.1. The Gauss-functions $g(t,\alpha) = \frac{1}{\sqrt{\pi\alpha}} e^{-t^2/\alpha}$ for three different values of α. The smaller α, the higher the peak

by the function

$$\frac{1}{\sqrt{\pi\alpha}} e^{-t^2/\alpha} , \tag{4.1}$$

where the factor in front of the exponential function serves for the normalization of (4.1)

$$\int_{-\infty}^{+\infty} \frac{1}{\sqrt{\pi\alpha}} e^{-t^2/\alpha} dt = 1 , \tag{4.2}$$

i.e. the area under the pulse is equal to unity. Now let us try to make this pulse shorter and shorter. Mathematically this is achieved by letting

$$\alpha \to 0 . \tag{4.3}$$

Clearly, then the pulse becomes so short that outside of $t = 0$ it vanishes, whereas at $t = 0$ it still remains normalized according to (4.2). This leads us to the definition of the δ-function

$$\delta(t) = 0 \quad \text{for} \quad t \neq 0 , \tag{4.4}$$

$$\int_{-\epsilon}^{\epsilon} \delta(t) = 1 , \tag{4.5}$$

where ϵ may be arbitrarily small (Fig. 4.2). Whereas it does not make sense to put $\alpha = 0$ in (4.1), the δ-function defined by the limit (4.3) does make sense, though only under an integral. Let us study a few properties of that function. Instead of centering the pulse around $t = 0$, we can center it around any other time t_0 (Fig. 4.3) so that (4.4) and (4.5) are transformed into

$$\delta(t - t_0) = 0 \quad \text{for} \quad t \neq t_0 , \tag{4.6}$$

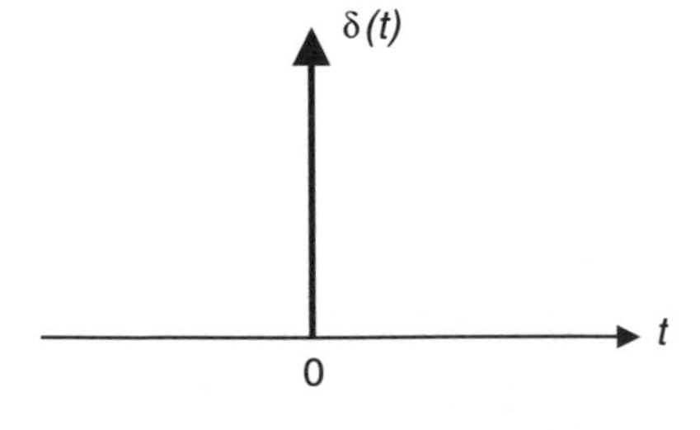

Fig. 4.2. δ-function located at $t = 0$

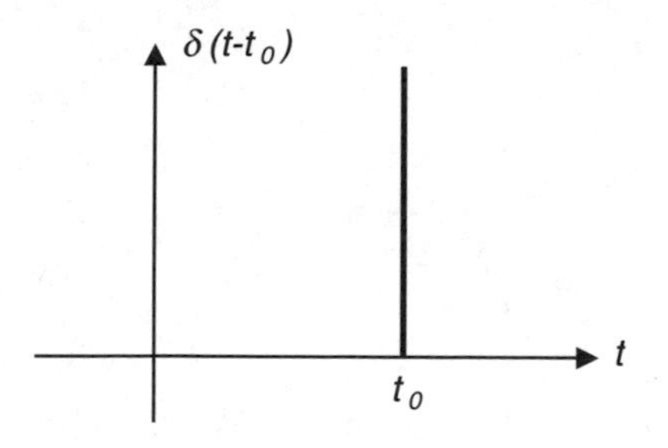

Fig. 4.3. δ-function located at $t = t_0$

and

$$\int_{t_0-\epsilon}^{t_0+\epsilon} \delta(t - t_0)dt = 1 . \tag{4.7}$$

To show the equivalence of (4.6) and (4.7) with (4.4) and (4.5), we just have to introduce a new variable

$$t - t_0 = s , \tag{4.8}$$

and change the limits of the integral

$$t = t_0 \pm \epsilon \tag{4.9}$$

accordingly, which yields

$$s = \pm\epsilon , \tag{4.10}$$

so that we finally transform (4.7)) into

$$\int_{-\epsilon}^{\epsilon} \delta(s)ds = 1 , \tag{4.11}$$

which is, of course, identical with (4.5). The particular properties of the δ-function lead to a number of quite useful formulas. If $h(t)$ is a continuous function, then (Fig. 4.4)

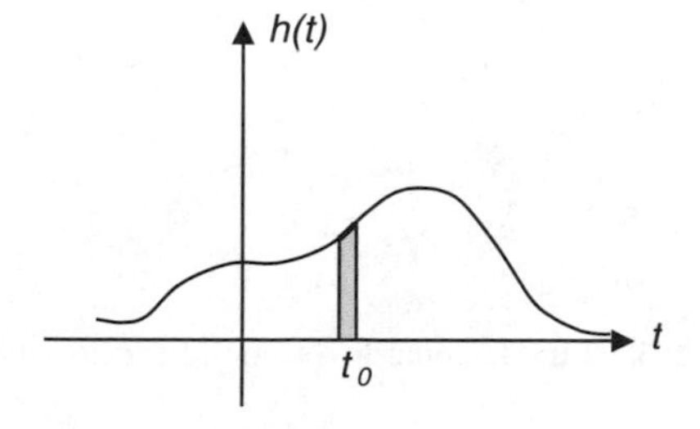

Fig. 4.4. Illustration of (4.13)

$$\int_{-\epsilon}^{\epsilon} h(t)\delta(t)dt = h(0)\,, \tag{4.12}$$

$$\int_{t_0-\epsilon}^{t_0+\epsilon} h(t)\delta(t-t_0)dt = h(t_0)\,. \tag{4.13}$$

We leave it to the reader as an exercise to derive these formulas. We can also define temporal derivatives of the δ-function, again by means of an integral. (In this book, in most cases we abbreviate temporal derivatives, such as dh/dt, by $\dot{h}$.) The following formula holds

$$\int_{-\epsilon}^{\epsilon} h(t)\dot{\delta}(t)dt = -\int_{-\epsilon}^{\epsilon} \dot{h}(t)\delta(t)dt = -\dot{h}(0)\,, \tag{4.14}$$

which may be easily proved by a partial integration of the l.h.s. of (4.14). Another interesting property of the δ-function results when we consider the integral

$$H(T) = \int_{-\infty}^{T} \delta(t)dt = \begin{cases} 0 \text{ for } T < 0 \\ 1 \text{ for } T > 0 \end{cases}, \tag{4.15}$$

which we may supplement if needed by the definition

$$H(T) = \frac{1}{2} \text{ for } T = 0\,. \tag{4.16}$$

The function $H(T)$ defined by (4.15) and (4.16) is a step function that is also called Heaviside function (Fig. 4.5).

Let us now study the property of a δ-function in which its argument t is replaced by ct

$$\int_{-\epsilon}^{\epsilon} \delta(ct)dt\,. \tag{4.17}$$

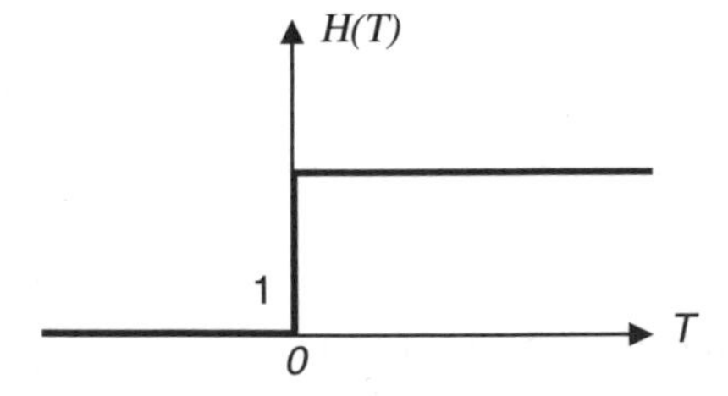

Fig. 4.5. The Heaviside (step-) function $H(T)$

Making the substitution

$$ct = s \tag{4.18}$$

so that

$$dt = \frac{1}{c} ds \,, \tag{4.19}$$

and replacing the boundaries of the integral

$$t = \pm\epsilon \tag{4.20}$$

by

$$s = \pm c\epsilon \,, \tag{4.21}$$

we readily obtain

$$\int_{-c\epsilon}^{c\epsilon} \delta(s) \frac{1}{c} ds = \frac{1}{c} \,. \tag{4.22}$$

It is now a simple exercise to combine the properties (4.6) and (4.17) to study

$$\int_{T_1}^{T_2} \delta(ct - t_0) dt \,, \tag{4.23}$$

where we will assume $c > 0$. Then we use the new variable s by means of

$$ct - t_0 = s \,. \tag{4.24}$$

It then follows that

$$\int_{-\infty}^{T} \delta(ct - t_0) dt = \begin{cases} 0 & \text{for } T < t_0/c \\ 1/c & \text{for } T > t_0/c \\ 1/(2c) & \text{for } T = t_0/c \,. \end{cases} \tag{4.25}$$

The δ-functions in (4.6), (4.17) and (4.23) are special cases of a δ-function, which itself depends on a function of time. We call this function $\phi(t)$, because in our neural models it will have the meaning of a phase (see Sect. 4.9 and Fig. 4.6). Let us study

$$\int_{t_1}^{T} \delta(\phi(t)) dt \,. \tag{4.26}$$

We assume that

$$\text{for} \quad t_1 \le t \le T \quad \text{the only zero of } \phi(t) \text{ is at } \; t = t_0 \,, \tag{4.27}$$

as also

$$\frac{d\phi(t_0)}{dt} \equiv \dot{\phi}(t_0) \neq 0 \tag{4.28}$$

and

$$\dot{\phi}(t_0) > 0 \,. \tag{4.29}$$

These are actually conditions that are fulfilled in our later applications. How to proceed further is now quite similar to our above examples (4.7), (4.17) and (4.23). We introduce a new variable by means of

$$s = \phi(t) \tag{4.30}$$

so that

$$ds = \dot{\phi}(t)dt \,. \tag{4.31}$$

We further obtain the correspondence

$$\phi(t_0) \leftrightarrow s = 0 \,. \tag{4.32}$$

By means of (4.30) and (4.31), (4.26) is tranformed into

$$\int\limits_{\phi(t_1)}^{\phi(T)} \delta(s) \frac{1}{\dot{\phi}(t(s))} ds \,, \tag{4.33}$$

where we note

$$\dot{\phi}(t(0)) = \dot{\phi}(t_0) \,. \tag{4.34}$$

We further assume (which actually follows from (4.27)–(4.29))

$$\phi(t_1) < \phi(T) \,, \tag{4.35}$$

so that our final result for (4.26) reads

$$\int\limits_{t_1}^{T} \delta(\phi(t))dt = \begin{cases} 0 & \text{for } T < t_0 \,, \\ 1/\dot{\phi}(t_0) & \quad T > t_0 \end{cases} \tag{4.36}$$

and

$$= \frac{1}{2} \cdot \frac{1}{\dot{\phi}(t_0)} \quad \text{for } T = t_0 \,. \tag{4.37}$$

Because of the occurrence of the factor $\dot{\phi}$ in (4.36), the δ-function depending on $\phi(t)$ is rather diffcult to handle in our later applications. For this reason, we define a new function

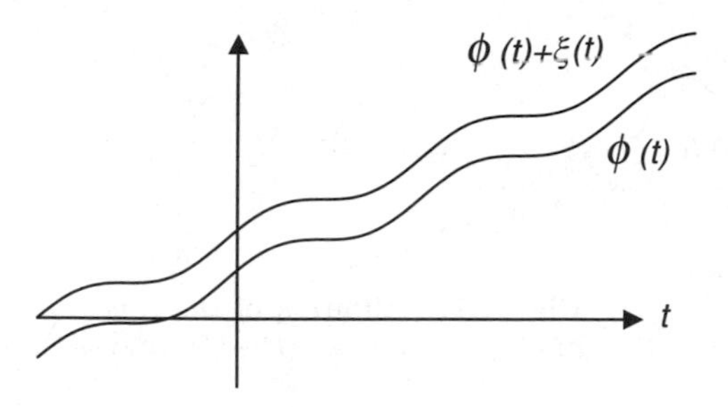

Fig. 4.6. Example of variation of the phases $\phi(t)$ and $\phi(t) + \xi(t)$ as functions of time t

$$f(t) = \delta(\phi(t))\dot{\phi}(t) , \tag{4.38}$$

which, when replacing the original δ-function, yields

$$\int_{t_0-\epsilon}^{t_0+\epsilon} f(t)dt \equiv \int_{t_0-\epsilon}^{t_0+\epsilon} \delta(\phi(t))\dot{\phi}(t)dt = 1 , \tag{4.39}$$

and which, like the δ-function, represents a spike at $t = t_0$. In our neural models (see Chaps. 5 and 6), we will use (4.38) instead of the δ-function.

4.2 Perturbed Step Functions

As we will see later in our neural models, the generation of pulses may be perturbed so that their phases are shifted, or in other words so that $\phi(t)$ is replaced by $\phi(t) + \xi(t)$ (Fig. 4.6). We then will have to study how big the change of the step function (4.15) (Heaviside function) is when we use the *spike function* (4.38) instead of the δ-function in (4.15). To this end, we treat the expression

$$G(T) = \int_{t_1}^{T} \delta(\phi(t) + \xi(t))(\dot{\phi}(t) + \dot{\xi}(t))dt - \int_{t_1}^{T} \delta(\phi(t))\dot{\phi}(t)dt . \tag{4.40}$$

Repeating all the steps we have done before with respect to the two integrals in (4.40), we readily obtain

$$G(T) = \int_{\phi(t_1)+\xi(t_1)}^{\phi(T)+\xi(T)} \delta(s)ds - \int_{\phi(t_1)}^{\phi(T)} \delta(s)ds \equiv I_1 - I_2 . \tag{4.41}$$

By means of Figs. 4.7–4.10 we can easily deduce the properties of $G(T)$. Let us start with the second integral in (4.41) and call it I_2. We assume that its lower limit is negative. Then I_2 is a step function with its step at $s = \phi(T) = 0$ (Fig. 4.8).

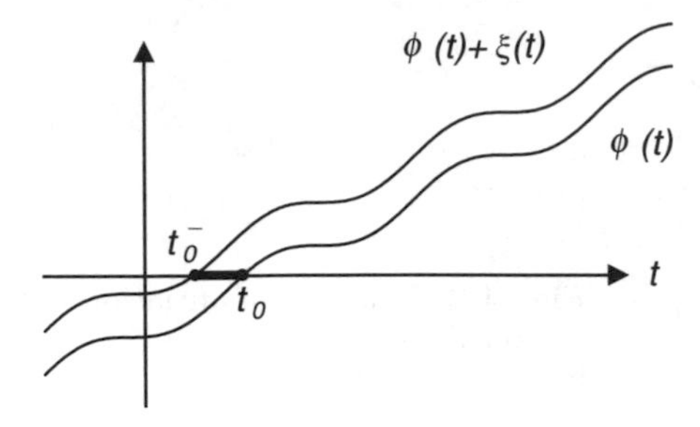

Fig. 4.7. Definition of the times t_0^- and t_0 as zeros of $\phi(t)+\xi(t)$ and $\phi(t)$, respectively

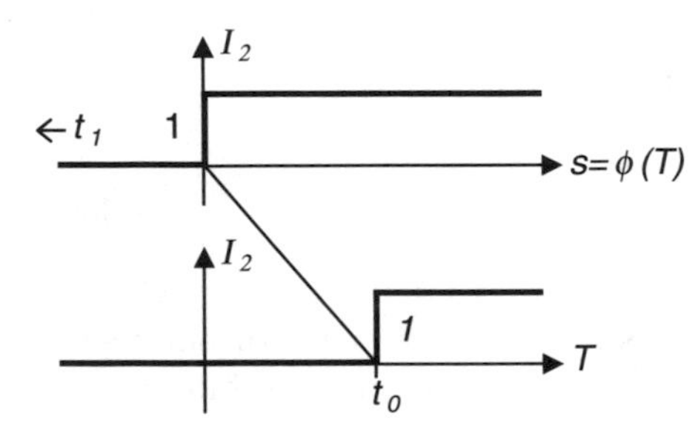

Fig. 4.8. *Upper part*: I_2 (see (4.41)) as a function of $s = \phi(T)$; *Lower part*: I_2 as a function of T

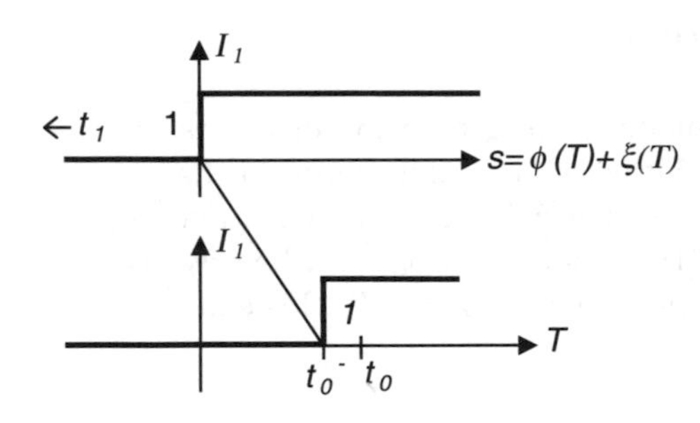

Fig. 4.9. *Upper part*: I_1 (see (4.41)) as a function of $s = \phi(T) + \xi(T)$; *Lower part*: I_1 as a function of T. Note the difference between the lower parts of Figs. 4.8 and 4.9

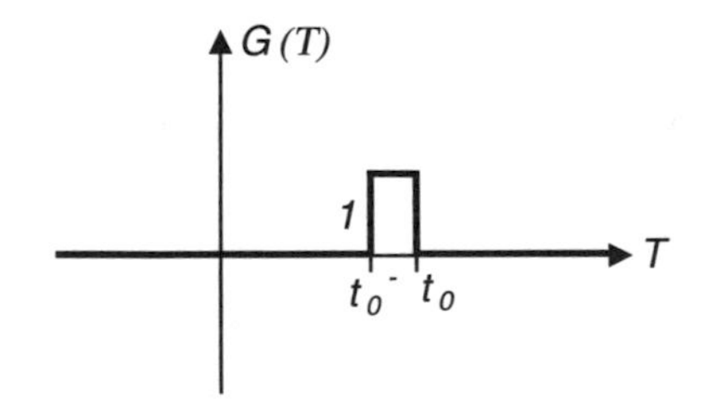

Fig. 4.10. $G(T) = I_1 - I_2$ results from the subtraction of the ordinate of Fig. 4.8 from that of Fig. 4.9

When we use the variable T instead of s, the step occurs at $T = t_0$, whereby

$$\phi(t_0) = 0 \tag{4.42}$$

(see also (4.27)). A similar consideration holds for the first integral in (4.40), I_1. Its plot against the variable $s = \phi(T) + \xi(T)$ is shown in the upper part of Fig. 4.9; the jump occurs at $s = 0$. But when we use the variable T, the jump occurs at t_0^-, where

$$\phi(t_0^-) + \xi(t_0^-) = 0\,. \tag{4.43}$$

The function $G(T)$, being the difference between I_1 and I_2, can be derived from the lower parts of Figs. 4.7 and 4.8 respectively, and is represented in Fig. 4.10. Thus

$$G(T) = \begin{cases} 0 \text{ for } T < t_0^- \\ 1 \text{ for } t_0^- < T < t_0\,. \\ 0 \text{ for } T > t_0 \end{cases} \tag{4.44}$$

We now turn to the explicit determination of t_0^- (see Fig. 4.7). This time t_0^- is defined by (4.43), whereas t_0 is defined by (4.42). In the following we will utilize the fact that in our practical applications the difference between t_0 and t_0^- is small. Subtracting (4.42) from (4.43) yields

$$\phi(t_0^-) - \phi(t_0) + \xi(t_0^-) = 0\,, \tag{4.45}$$

and using the Taylor expansion of $\phi(t)$ up to the second term

$$(t_0^- - t_0)\dot{\phi}(t_0) = -\xi(t_0^-)\,. \tag{4.46}$$

Resolving (4.46) for the time difference yields

$$(t_0 - t_0^-) = \xi(t_0^-)/\dot{\phi}(t_0)\,. \tag{4.47}$$

If $\xi(t)$ varies smoothly, we may use the further approximation

$$\xi(t_0^-) = \xi(t_0) + (t_0^- - t_0)\dot{\xi}(t_0)\,. \tag{4.48}$$

Inserting this approximation into (4.47) yields

$$(t_0 - t_0^-)(1 - \dot{\xi}(t_0)/\dot{\phi}(t_0)) = \xi(t_0)/\dot{\phi}(t_0)\,. \tag{4.49}$$

Provided

$$|\, \dot{\xi}(t_0)/\dot{\phi}(t_0) \,| \ll 1 \tag{4.50}$$

holds, we arrive at our final result

$$(t_0 - t_0^-) = \xi(t_0)/\dot{\phi}(t_0)\,. \tag{4.51}$$

Let us now consider what happens when we perform an integral over the function $G(T)$ over the interval

$$T_1 \le T \le T_2\,. \tag{4.52}$$

In principle, we must differentiate between different cases, e.g.

$$T_1 < t_0^- < T_2 < t_0 \,, \tag{4.53}$$

and further time-sequences. For many practical purposes it suffices, however, to treat the case

$$T_1 < t_0^- \leq t_0 < T_2 \,. \tag{4.54}$$

Using (4.44), we obtain

$$\int_{T_1}^{T_2} G(T)dT = \begin{cases} 0 & \text{for } T_2 < t_0^- \leq t_0 \\ \int_{t_0^-}^{t_0} dT = t_0^- - t_0 & \text{for } T_1 < t_0^- \leq t_0 < T_2 \,. \\ 0 & \text{for } T_1 > t_0^-, t_0 \end{cases} \tag{4.55}$$

This result reminds us of the property of a δ-function with one basic difference. Whereas the δ-function vanishes everywhere except at the point where its argument vanishes, the function $G(T)$ vanishes only outside the interval $t_0^- ... t_0$, whereas within this interval it acquires the value unity. If the interval $t_0 - t_0^-$ is small, for many practical purposes we may replace the function $G(T)$ by a δ-function, where we have to add a weight factor that yields the same area. In other words, we replace the l.h.s. of (4.55) by

$$\int_{T_1}^{T_2} (t_0 - t_0^-)\delta(T - t_0)dT = \begin{cases} 0 & \text{for } T_2 < t_0 \\ (t_0 - t_0^-) & \text{for } T_1 < t_0 < T_2 \\ 0 & \text{for } T_1 > t_0 \end{cases} \tag{4.56}$$

with the properties presented on the r.h.s. of this equation. If $t_0 - t_0^-$ is small and the function $h(T)$ varies only a little in this interval, we may use the following approximation

$$\int_{T_1}^{T_2} h(T)G(T)dT \approx h(t_0) \int_{T_1}^{T_2} G(T)dT \,. \tag{4.57}$$

This means, jointly with (4.55), that $G(T)$ has practically the properties of the δ-function!

4.3 Some More Technical Considerations*

Readers who are not interested in mathematical details, can skip this section. In some practical applications that will follow in later chapters, it turns out that while

$$\phi(t) \quad \text{is continuous at} \quad t = 0 \tag{4.58}$$

we must observe that

$$\dot{\phi}(t) \quad \text{is discontinuous at} \quad t = 0\,. \tag{4.59}$$

This leads us to the question of how to evaluate

$$\delta(\phi(t))\dot{\phi}(t)? \tag{4.60}$$

To this end, we recall that the δ-function can be conceived as a specific limit $\alpha \to 0$ of

$$\sqrt{\frac{1}{\pi\alpha}} e^{-s^2/\alpha}\,. \tag{4.61}$$

Therefore we consider

$$\sqrt{\frac{1}{\pi\alpha}} e^{-g^2(t)/\alpha} \dot{\phi}(t) : \text{ for} \begin{matrix} t < 0 \\ t > 0 \end{matrix}\,. \tag{4.62}$$

Because we have to distinguish between positive and negative times, we perform the following split

$$\int\limits_{-\infty}^{+\infty} ...ds = \int\limits_{-\infty}^{0-} ..ds + \int\limits_{0+}^{\infty} ..ds = \frac{1}{2} + \frac{1}{2} = 1\,. \tag{4.63}$$

Inserting (4.61) into the two parts yields the result as indicated on the r.h.s. of (4.63). In complete analogy, we may also perform the following split with the obvious result

$$\int\limits_{-\epsilon}^{\epsilon} \delta(s)ds = \int\limits_{-\epsilon}^{0-} \delta(s)ds + \int\limits_{0+}^{\epsilon} \delta(s)ds = \frac{1}{2} + \frac{1}{2} = 1\,. \tag{4.64}$$

With these precautions in mind, we can repeat the above evaluations of Sect. 4.2. We need, however, some precautions in order to evaluate t_0^-, because in that evaluation the derivative $\dot{\phi}$ was involved. Repeating the above steps for the two regions for t, or correspondingly ξ, we obtain

$$(t_0^- - t_0) = -\xi(t_0^-) \cdot \begin{cases} \dot{\phi}_-(t_0)^{-1}, \, \xi > 0 \\ \dot{\phi}_+(t_0)^{-1}, \, \xi < 0 \end{cases}. \tag{4.65}$$

We leave it as an exercise to the reader to derive this result. In order to find a formula that combines the two cases of (4.65), we introduce the identities

$$\frac{1}{2}(\xi + \mid \xi \mid) = \begin{cases} \xi \text{ for } \xi > 0 \\ 0 \text{ for } \xi < 0 \end{cases} \tag{4.66}$$

and

$$\frac{1}{2}(\xi - \mid \xi \mid) = \begin{cases} 0 \text{ for } \xi > 0 \\ \xi \text{ for } \xi < 0 \end{cases}. \tag{4.67}$$

Thus we can write (4.65) in the form

$$t_0^- - t_0 = -\dot{\phi}_-(t_0)^{-1}\left(\frac{1}{2}(\xi + \mid \xi \mid)\right) + \dot{\phi}_+(t_0)^{-1}\left(\frac{1}{2}(\xi - \mid \xi \mid)\right) \tag{4.68}$$

that can be rearranged to yield

$$\begin{aligned} t_0^- - t_0 = {} & \xi(t_0^-)\tfrac{1}{2}\left(\dot{\phi}_-(t_0)^{-1} + \dot{\phi}_+(t_0)^{-1}\right) \\ & + \mid \xi(t_0^-) \mid \tfrac{1}{2}\left(\dot{\phi}_-(t_0)^{-1} - \dot{\phi}_+(t_0)^{-1}\right). \end{aligned} \tag{4.69}$$

This relation can be considerably simplified if the second term in (4.69) is much smaller than the first one, i.e. provided

$$\frac{\dot{\phi}_+ - \dot{\phi}_-}{\dot{\phi}_+ + \dot{\phi}_-} \ll 1 \tag{4.70}$$

holds. We then obtain our final result

$$t_0^- - t_0 = \xi(t_0^-)\frac{1}{2}\left(\dot{\phi}_-(t_0)^{-1} + \dot{\phi}_+(t_0)^{-1}\right). \tag{4.71}$$

4.4 Kicks

Later in this book, we want to model the conversion of axonal spikes into dendritic currents. At the molecular level, this is quite a complicated process. But as discussed in Chap. 1, we want to study brain functions at and above the level of neurons. What happens in a dendrite before an axonal pulse arrives at a synapse? Clearly, there is no dendritic current; then the pulse generates a current, which finally is damped out. To model this phenomenon, we invoke the mechanical example of a soccer ball that is kicked by a soccer player and rolls over grass, whereby its motion will be slowed down. In this case, it is rather obvious how to describe the whole process. Our starting point is Newton's law according to which the velocity v of a particle with mass m changes according to the equation

$$m\frac{dv}{dt} = \text{force}. \tag{4.72}$$

In order to get rid of superfluous constants, at least for the time being, we put $m = 1$. The force on the r.h.s. consists of the damping force of the grass that we assume to be proportional to the velocity v and the individual kick of the soccer player. Because the kick lasts only for a short time, but is very strong, we describe it by means of a δ-function. In this way, we formulate the equation of motion as

$$\frac{dv(t)}{dt} = -\gamma v(t) + s\delta(t - \sigma), \tag{4.73}$$

where γ is the damping constant and s the strength of the kick. We assume that at an initial time $t_0 < \sigma$, the velocity of the soccer ball is zero

$$v(t_0) = 0\,. \tag{4.74}$$

For the time interval $t_0 \le t < \sigma$, i.e. until the kick happens, the soccer ball obeys the equation

$$\frac{dv(t)}{dt} = -\gamma v(t)\,. \tag{4.75}$$

Because it is initially at rest, it will remain so

$$v(t) = 0\,. \tag{4.76}$$

Now the exciting problem arises, namely to describe the effect of the kick on the soccer ball's motion. Since the definition of the δ-function implies that it appears under an integral, we integrate both sides of (4.73) over a short time interval close to $t = \sigma$

$$\int_{\sigma-\epsilon}^{\sigma+\epsilon} \frac{dv(t)}{dt} dt = \int_{\sigma-\epsilon}^{\sigma+\epsilon} -\gamma v(t) dt + \int_{\sigma-\epsilon}^{\sigma+\epsilon} s\delta(t-\sigma) dt\,. \tag{4.77}$$

Since integration is just the inverse process to differentiation, we may evaluate the l.h.s. immediately. Assuming that ϵ tends to zero, the first integral on the r.h.s. will vanish, while the second integral just yields s, because of the property of the δ-function. Thus (4.77) is transformed into

$$v(\sigma+\epsilon) - v(\sigma-\epsilon) = s\,, \tag{4.78}$$

and because before the kick the soccer ball was at rest, we obtain

$$v(\sigma+\epsilon) = s\,. \tag{4.79}$$

Now consider the time $t \ge \sigma$. Then (4.73) reads again (4.75), which possesses the general solution

$$v(t) = ae^{-\gamma t}\,, \tag{4.80}$$

as one can immediately verify by inserting (4.80) into (4.75). a is a constant that must be chosen appropriately, namely such that at time $t = \sigma$ the velocity becomes (4.79). This can be achieved by putting

$$a = se^{\gamma\sigma} \tag{4.81}$$

so that the final solution to (4.75) with the initial condition (4.79) reads

$$v(t) = se^{-\gamma(t-\sigma)}, \quad t \ge \sigma\,. \tag{4.82}$$

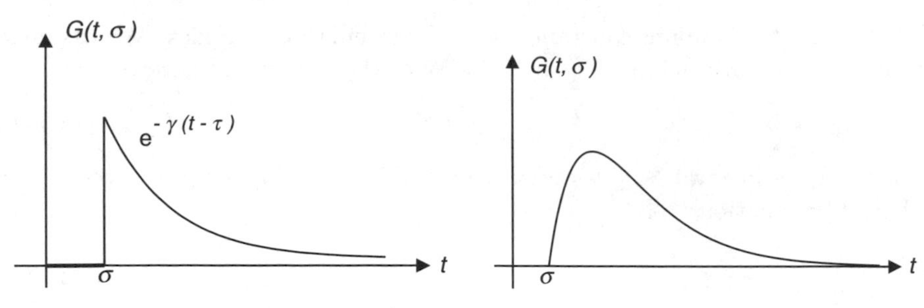

Fig. 4.11. The Green's function (4.83) **Fig. 4.12.** The Green's function (4.84)

Setting $s = 1$, we obtain an important special case, namely that of a unit kick. In this case, v is denoted by $G(t, \sigma)$ and is called a *Green's function.* Obviously, this function is defined by

$$G(t, \sigma) = \begin{cases} 0 & \text{for } t < \sigma \\ e^{-\gamma(t-\sigma)} & \text{for } t \geq \sigma \end{cases} . \tag{4.83}$$

In the context of our book, where we are concerned with the generation of a dendritic current by a δ-pulse (spike), (4.83) provides us with a first answer: A current is immediately generated and then damped (Fig. 4.11). We will use this approach in Chaps. 5 and 6. A more refined approach takes into account that the current first increases linearly with time until it reaches its maximum after which it drops exponentially.

This behavior, shown in Fig. 4.12, is represented by

$$G(t, \sigma) = \begin{cases} 0 & \text{for } t < \sigma \\ (t-\sigma)e^{-\gamma(t-\sigma)} & \text{for } t \geq \sigma \end{cases} . \tag{4.84}$$

Using the techniques we just have learnt, we can easily show that (4.84) is the Green's function belonging to

$$\left(\frac{d}{dt} + \gamma \right)^2 G(t, \sigma) = \delta(t - \sigma) \tag{4.85}$$

(see the exercise below). In the literature, often the *normalization factor* γ^2 is added. The resulting function is called (Rall's) α-function (where formally α is used instead of γ). We will use this function in Chap. 7.

Exercise. Show that (4.84) satisfies (4.85).
Hint: Write

$\left(\frac{d}{dt} + \gamma\right)^2 G \equiv \frac{d^2G}{dt^2} + 2\gamma \frac{dG}{dt} + \gamma^2 G = \delta(t - \sigma)$.

Convince yourself that (4.84) satisfies this equation for $t < \sigma$ and $t > \sigma$. For $\sigma - \epsilon < t < \sigma + \epsilon$ integrate both sides of (4.83) over this interval. Show that $\int_{-\infty}^{+\infty} \alpha^2 G(t, \sigma) dt = 1$.

4.5 Many Kicks

Let us start with the example of two consecutive kicks in which case the force exerted by the soccer player on the ball can be written in the form

$$F(t) = s_1\delta(t - \sigma_1) + s_2\delta(t - \sigma_2)\,, \tag{4.86}$$

where the kicks are performed at times σ_1 and σ_2 with strengths s_1 and s_2, respectively. The equation of motion then reads

$$\frac{dv(t)}{dt} = -\gamma v(t) + s_1\delta(t - \sigma_1) + s_2\delta(t - \sigma_2)\,. \tag{4.87}$$

Because (4.87) is linear, for its solution we make the hypothesis

$$v(t) = v_1(t) + v_2(t)\,. \tag{4.88}$$

Using the indices $j = 1, 2$, we require that v_j obeys the equation

$$\frac{dv_j}{dt} = \gamma v_j(t) + s_j\delta(t - \sigma_j)\,. \tag{4.89}$$

We know already that its solution can be expressed by means of the Green's function (4.83) in the form

$$v_j(t) = s_j G(t, \sigma_j)\,. \tag{4.90}$$

By use of (4.88), we obtain

$$v(t) = s_1 G(t, \sigma_1) + s_2 G(t, \sigma_2) \tag{4.91}$$

as the final solution to (4.87). Now it is an obvious task to treat many kicks, in which case the forces are composed of many contributions in the form

$$F(t) = \sum_{j=1}^{N} s_j\delta(t - \sigma_j)\,, \tag{4.92}$$

and by analogy to (4.91) we obtain the final result (Fig. 4.13)

$$v(t) = \sum_{j=1}^{N} s_j G(t, \sigma_j)\,. \tag{4.93}$$

Let us make a final generalization in which case the kicks are continuously exerted on the soccer ball. In this case, the force can be written as an integral in the form

$$F(t) = \int_{t_0}^{T} s(\sigma)\delta(t - \sigma)d\sigma \equiv \int_{t_0}^{T} F(\sigma)\delta(t - \sigma)d\sigma\,, \tag{4.94}$$

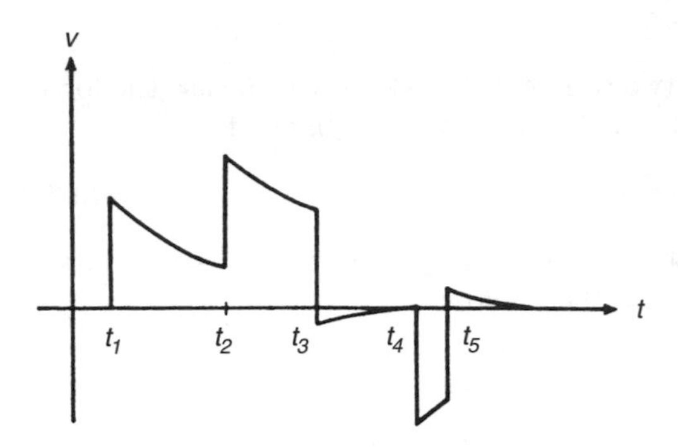

Fig. 4.13. The velocity $v(t)$ caused by δ-pushes at times $\sigma_1, \sigma_2, ...$ and damping

where we leave it as a little exercise to the reader to verify the second representation. In the following we will assume that T goes to infinity when the whole process is considered for an arbitrarily long time. The continuous version of the solution (4.93) obviously reads

$$v(t) = \int_{t_0}^{T} F(\sigma)G(t, \sigma)d\sigma \,. \tag{4.95}$$

Now it is useful to use the explicit form of the Green's function (4.83), which allows us to cast (4.95) into the final form

$$v(t) = \int_{t_0}^{t} F(\sigma)e^{-\gamma(t-\sigma)}d\sigma \,. \tag{4.96}$$

We leave it as an exercise to the reader to verify that the solution (4.96) must be complemented by a solution of the homogeneous equation, i.e. with $F = 0$ if at the initial time t_0 the velocity did not vanish.

4.6 Random Kicks or a Look at Soccer Games

Quite often we observe in soccer games that the players kick the ball entirely at random as far as both the direction and timing are concerned. The reader may wonder what the study of soccer games will have to do with our task of modeling neurons, but I hope that will become evident in a few moments. We now denote the times at which the kicks occur by t_j and indicate their direction in a one-dimensional game by $(\pm 1)_j$, where the choice of the plus or minus sign is random, for instance, it is determined by throwing a coin. Thus the kicking force can be written in the form

$$F(t) = s \sum_j \delta(t - t_j)(\pm 1)_j \,, \tag{4.97}$$

where for simplicity we assume that all kicks have the same strength. When we observe many games, then we may perform an average over all these different performances. Let us denote this average by $< ... >$. To treat the averaging procedure in detail, we consider

$$< F(t) >= s < \sum_j \delta(t - t_j)(\pm 1)_j > . \tag{4.98}$$

Since the direction of the kicks is assumed to be independent of the time at which the kicks happen, we may split (4.98) into the product

$$(4.98) = s < \sum_j \delta(t - t_j) >< (\pm 1)_j > . \tag{4.99}$$

Because the kicks are assumed to happen with equal frequency in the positive and negative directions, we obtain a cancellation

$$< (\pm 1)_j >= 0 . \tag{4.100}$$

Thus the average of $F(t)$ (4.98) vanishes,

$$< F(t) >= 0 . \tag{4.101}$$

In order to characterize the strength of the force (4.97), we consider a quadratic expression in F. For later purposes we evaluate the following *correlation function for two times* t, t', where by means of (4.97) we obtain

$$< F(t)F(t') >= s^2 < \sum_j \delta(t - t_j)(\pm 1)_j \sum_k \delta(t' - t_k)(\pm 1)_k > , \tag{4.102}$$

which in analogy with (4.98) can be split into a term that contains the directions of the kicks and another term. In this way, we are led to study

$$< (\pm 1)_j (\pm 1)_k > = < (\pm 1)_j >< (\pm 1)_k > \quad \text{for} \quad j \neq k , \tag{4.103}$$

where we could split the l.h.s. into a product because of the statistical independence of the kicks. Because of (4.100), the expressions in (4.103) vanish. On the other hand, we obviously obtain

$$< (\pm 1)_j (\pm 1)_k > = < 1 > = 1 \quad \text{for} \quad j = k . \tag{4.104}$$

Thus the sums over j and k in (4.102) reduce to a single sum and we obtain

$$< F(t)F(t') >= s^2 < \sum_j \delta(t - t_j)\delta(t' - t_j) > \tag{4.105}$$

instead of (4.102). Usually the r.h.s. of (4.105) is evaluated by assuming a so-called Poisson process for the times of the kicks. For our purposes it is sufficient to evaluate the r.h.s. by taking an average over the time T of a game, multiplied by the number N of kicks during T. Then we obtain

$$< ... >= N\frac{1}{T}\int_0^T \delta(t-\tau)\delta(t'-\tau)d\tau\,, \tag{4.106}$$

which can be evaluated to yield

$$< ... >= \frac{N}{T}\delta(t-t')\,. \tag{4.107}$$

Readers who are afraid of multiplying δ-functions and integrating over their product (at different times!) may be reminded that δ-functions can be approximated for instance by Gaussian integrals and then the result (4.107) can be easily verified. Putting $T/N = t_0$, which can be interpreted as a mean free time between kicks, and putting

$$s^2/t_0 = Q\,, \tag{4.108}$$

our final result reads

$$< F(t)F(t') >= Q\delta(t-t_0)\,. \tag{4.109}$$

4.7 Noise Is Inevitable. Brownian Motion and the Langevin Equation

The model we have developed above may seem to be rather arbitrary. In this section we want to show that it has an important application to the physical world. The phenomenon we have in mind is Brownian motion. When a particle is immersed in a fluid, the velocity of this particle if slowed down by a force proportional to the velocity of this particle. When one studies the motion of such a particle under a microscope in more detail, one realizes that this particle undergoes a zig–zag motion. This effect was first observed by the biologist Brown. The reason for zig–zag motion is this: The particle under consideration is steadily pushed by the much smaller particles of the liquid in a random way. Let us describe the whole process from a somewhat more abstract viewpoint. Then we deal with the behavior of a system (namely the particle), which is coupled to a heat bath or reservoir (namely the liquid). The heat bath has two effects:

1. it decelerates the mean motion of the particle; and
2. it causes statistical fluctuations.

In the context of the previous section, the modelling of the whole process is rather obvious and was done in this way first by Langevin by writing down the following equation of motion for the particle, where we put $m = 1$. F is denoted as a fluctuating force and has the following properties:

1. its statistical average, i.e. (4.98), vanishes; and
2. its correlation function is given by (4.109).

For what follows, all we need are these two properties and the form of the solution to

$$\frac{dv(t)}{dt} = -\gamma v(t) + F(t) , \tag{4.110}$$

which, quite generally, is given by (4.96). Let us determine the average velocity $< v(t) >$, where the average is taken over all games. Because the integration in (4.96) and this average are independent of each other, we may perform the average under the integral, which because of (4.101) vanishes. Thus,

$$< v(t) >= 0 , \tag{4.111}$$

i.e. on the average the velocity vanishes, which stems from the fact that both directions are possible and cancel each other. Thus in order to get a measure for the size of the velocity as well as of its change in the course of time, we consider the quadratic expression in

$$< (v(t) - v(t'))^2 >=< v^2(t) > + < v^2(t') > -2 < v(t)v(t') > . \tag{4.112}$$

As we will see later, the first terms are equal and even time-independent in the steady state. The important term of physical interest is the third expression in (4.112), which we now study. Using the explicit form of v, it reads

$$< v(t)v(t') > = < \int_{t_0}^{t} d\sigma \int_{t_0}^{t'} d\sigma' e^{-\gamma(t-\sigma)} e^{-\gamma(t'-\sigma')} F(\sigma)F(\sigma') > . \tag{4.113}$$

Taking the averaging procedure under the integral, using the property (4.109) and evaluating the integrals is now an easy matter (see exercise below). In the steady state, we may assume that $t+t' \to \infty$. In this case, the final result reads

$$< v(t)v(t') >= (Q/2\gamma)e^{-\gamma(t-t')} \tag{4.114}$$

and for equal times

$$< v(t)^2 >= \frac{Q}{2\gamma} . \tag{4.115}$$

So far we have put the mass $m = 1$; it is a simple matter to repeat all the steps with $m \neq 1$ starting from Newton's equation. In such a case, the final result reads

$$< v(t)^2 >= \frac{Q}{2\gamma m} . \tag{4.116}$$

Now the decisive step occurs, namely the connection with the real world. Multiplying (4.116) by $m/2$ on the l.h.s. we obtain the mean kinetic energy of the particle immersed in the liquid. Because of a fundamental law of thermodynamics, this energy is given by

$$\frac{m}{2} < v(t)^2 >= \frac{1}{2}kT\,, \tag{4.117}$$

where k is Boltzmann's constant and T is the absolute temperature. Comparing (4.116) and (4.117), we obtain the following fundamental result derived by Einstein

$$Q = 2\gamma kT\,. \tag{4.118}$$

It tells us that whenever there is damping, i.e. $\gamma \neq 0$, then there are fluctuations Q. In other words, fluctuations or noise are inevitable in any physical system, which, of course, also applies to neurons. This derivation is not limited to our mechanical example. It is quite universal. An electric counterpart was studied experimentally by Johnson and theoretically by Nyquist. In a resistor, the electric field E fluctuates with a correlation function given by

$$< E(t)E(t') >= 2RkT\delta(t - t')\,, \tag{4.119}$$

where R is the resistance of the resistor. This is the simplest example of a so-called dissipation–fluctuation theorem. If we Fourier analyse (4.119) and denote the Fourier transform of $E(t)$ by $E(\omega)$, (4.119) can be transformed into

$$< E^2(\omega) >= RkT/\pi\,, \tag{4.120}$$

which is the *Nyquist theorem*. Quite clearly, when we want to formulate an appropriate theory of the action of neurons, we must take into account noise effects. But things are still more difficult and are surely not finally studied, because a neuron cannot be considered as a system in thermal equilibrium. Quite the contrary, it is a highly active system with noise still more pronounced.

Exercise. Evaluate (4.113) and verify (4.114).

4.8 Noise in Active Systems

4.8.1 Introductory Remarks

In the previous section we studied noise in systems that are in thermal equilibrium. As we noticed, the strength of the noise forces is quite universal, namely it depends only on the absolute temperature and on the damping constant, be it in a mechanical, electrical, or any other physical system. While thermal noise is certainly present in neural systems, e.g. in dendrites, another important aspect must be taken into account. In order to be able to process information, neurons must be active systems into which energy is pumped. This occurs, of course, by metabolism. In physics, there exists a detailed theory of noise in active systems. It has been developed in the context of lasers, nonlinear optics, and other devices, and is based on quantum theory. It would be far beyond the scope of this book to present such a theory here. Furthermore, a detailed theory of noise sources in neural systems is presently lacking. For both these reasons, we present some basic ideas on how the fluctuating forces can be formulated and what their properties are. For more details we must refer the reader to the references.

4.8.2 Two-State Systems

As we have seen in Chap. 2, basic processes in neurons go on in particular at the synapses and axons. In both cases, fundamentally we have to deal with ion transport through channels in membranes. Let us consider a membrane channel with an ion as a two-state system, where the ion may be either inside or outside the membrane. We indicate these states by the indices 1 (inside) and 2 (outside). While a precise treatment must be left to quantum theory, it may suffice here to outline the basic concepts. The occupation number of the states 1 or 2 will be denoted by n_1 and n_2, respectively. Because of transport through the membrane channel, these occupation numbers change in time and are described by the *rate equations*

$$\frac{dn_1}{dt} = -w_{21}n_1 + w_{12}n_2 + F_1(t) \tag{4.121}$$

and

$$\frac{dn_2}{dt} = w_{21}n_1 - w_{12}n_2 + F_2(t) \,. \tag{4.122}$$

Here the *transition rates* w_{12} and w_{21} do not only depend on the temperature T, but also on the voltage across the membrane

$$w_{12}(V)\,,\ w_{21}(V)\,. \tag{4.123}$$

As may be shown, in order to secure a quantum mechanically consistent description, the fluctuating forces $F_j(t), j = 1, 2$, must be included in the equations (4.121) and (4.122). As usual we may assume that the statistical averages vanish

$$< F_j(t) >= 0, \quad j = 1, 2\,, \tag{4.124}$$

because otherwise we would have to add a constant term to (4.121) and (4.122). When we assume as usual that the fluctuations have a short memory, compared to the ongoing processes, we may assume that they are δ-correlated

$$< F_j(t)F_k(t') >= Q_{jk}\delta(t-t') . \tag{4.125}$$

So far we have a formulation that is rather analogous to what we considered in the context of Brownian motion. The important difference, however, occurs with respect to the strengths of the fluctuations Q_{jk}. As it turns out, these quantities depend in a more intricate manner on the dynamics than in the case of Brownian motion, where only a damping constant appeared. In fact, as a detailed quantum mechanical calculation shows, the Qs are given by the relations

$$Q_{11} = Q_{22} = w_{21} < n_1 > + w_{12} < n_2 > \tag{4.126}$$

and

$$Q_{12} = Q_{21} = -w_{21} < n_1 > - w_{12} < n_2 > . \tag{4.127}$$

The average occupation numbers $< n_j >$ obey the equations

$$\frac{d < n_1 >}{dt} = -w_{21} < n_1 > + w_{12} < n_2 > \tag{4.128}$$

and

$$\frac{d < n_2 >}{dt} = w_{21} < n_1 > - w_{12} < n_2 > . \tag{4.129}$$

Thus we note that the fluctation strengths depend on the averaged occupation numbers, which may be time-dependent and which, as we have mentioned above, may depend on the voltage V across the membrane. Obviously, the noise sources of active systems are more complicated than those of systems in thermal equilibrium. The transition rates w_{21} and w_{12} may be either derived from a microscopic theory or can be introduced phenomenologically. In reality, a membrane contains many ion channels. Therefore, we want to study this case next.

4.8.3 Many Two-State Systems: Many Ion Channels

We distinguish the ion channels by an index μ so that we have to generalize the notation in the following obvious way

$$n_j \to n_{j,\mu}, \quad F_j \to F_{j,\mu}, \quad j = 1, 2 . \tag{4.130}$$

In practical applications we will be concerned with the properties of the total fluctuating forces that are defined by

$$F_j(t) = \sum_\mu F_{j,\mu} . \tag{4.131}$$

Because of (4.124), the average vanishes

$$< F_j(t) >= 0 \,. \tag{4.132}$$

In general, one may assume that the fluctuating forces of the individual channels are statistically independent. Therefore, if we form

$$< F_j(t)F_k(t') > = < \sum_\mu F_{j,\mu}(t) \sum_\nu F_{k,\nu}(t') > , \tag{4.133}$$

we readily obtain

$$(4.133) = \sum_\mu < F_{j,\mu}(t)F_{k,\mu}(t) > = \sum_\mu Q_{jk,\mu}\delta(t-t') , \tag{4.134}$$

where we made use of (4.125), then added to the fluctuation strength Q_{jk} of channel μ the corresponding index. In an obvious manner, the relation (4.126) generalizes to

$$Q_{jj,\mu} = w_{21} < n_1 >_\mu + w_{12} < n_2 >_\mu \,. \tag{4.135}$$

Introducing the total average number of the occupied states inside and outside the membrane, respectively, by

$$\sum_\mu < n_j >_\mu = N_j \,, \tag{4.136}$$

we obtain the relation

$$\sum_\mu Q_{jj,\mu} = w_{21}N_1 + w_{12}N_2 \,, \tag{4.137}$$

provided the transition rates of the channels under consideration can be considered as independent of the channel index μ. The relations (4.134) and (4.137) allow us to formulate our general result. The strength of the fluctuating forces is given by the sum of the number of total transitions per unit time out of the state 1 and the state 2, respectively. As we will see in a later chapter, the neural dynamics is determined by currents across the membrane. These currents are connected with the occupation numbers of the ion channels as introduced above by means of the relation

$$J = e\frac{dN_1}{dt} \,, \tag{4.138}$$

where e is the electric charge of the ion and J the electric current. With these currents also fluctuations are connected that can be determined by means of (4.138), (4.121), (4.122) and (4.131)–(4.137).

These hints may suffice here to give the reader an idea in which way fluctuating forces can be taken into account. For what follows, we need only the properties (4.132), (4.134) and (4.137).

Exercise. Determine $< J(t_1)J(t_2) >$.
Hint: Solve (4.121) and (4.122) with $n_1 + n_2 = 1$. Use the results analogous to (4.96) and (4.113).

4.9 The Concept of Phase

The occurrence of (more or less) regular spikes in an axon can be considered as a periodic event. Periodic or at least rhythmic events can also be observed in electric and magnetic fields of the brain, as measured by EEG and MEG, respectively. Rhythmic activity can be found in local field potentials. In order to quantitatively deal with rhythms, the concept of phase is indispensable. In this section, we will first present some elementary considerations, then proceed to a more sophisticated treatment that will allow us to extract phases from experimental data.

4.9.1 Some Elementary Considerations

Let us consider one of the simplest periodic events, namely the movement of a pendulum (Fig. 4.14). If we displace a pendulum (a swing) and if there is no friction, it will undergo a periodic movement forever. For not too high amplitudes, the displacement can be written in the form

$$\text{displacement} \quad x(t) = A \cos \omega t \,. \tag{4.139}$$

By differentiating x with respect to time, we obtain the velocity

$$\text{velocity} \quad v(t) \equiv \dot{x}(t) = -A\omega \sin \omega t \,. \tag{4.140}$$

In these relations, we use the circular frequency that is related to the period of one oscillation T by means of

$$\text{circular frequency} \quad \omega = 2\pi/T \,. \tag{4.141}$$

The displacement and velocity as described by (4.139) and (4.140) are chosen in such a way that the following initial conditions at time $t = 0$ are fulfilled

$$x(0) = A \tag{4.142}$$

and

$$\dot{x}(0) = 0 \,. \tag{4.143}$$

By means of

$$\omega t = \phi \,, \tag{4.144}$$

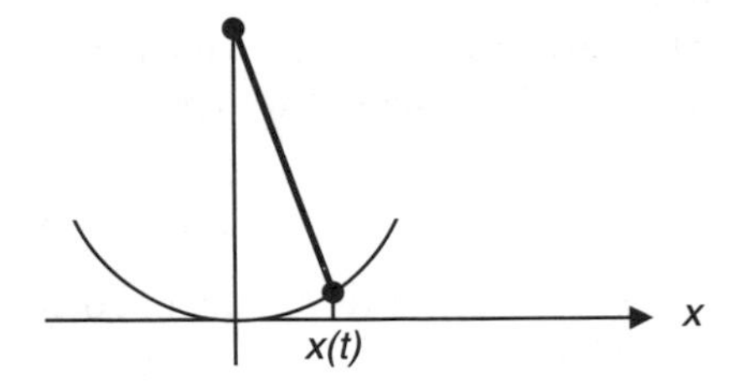

Fig. 4.14. Elongation $x(t)$ of a pendulum

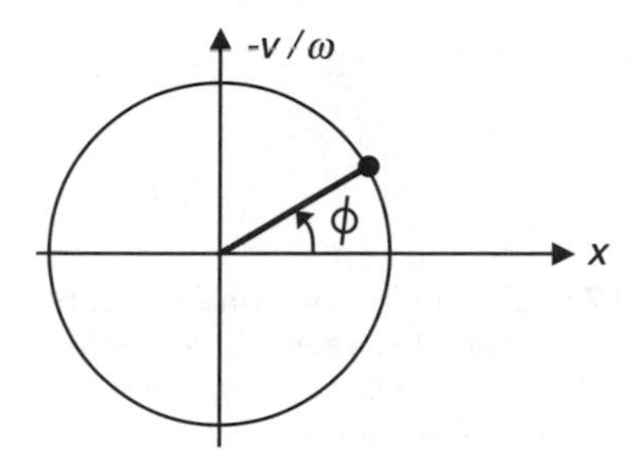

Fig. 4.15. Movement of the point $(x(t), -v(t)/\omega)$ on the orbit (here a circle) in the phase-plane. Note that the word "phase" has a double meaning: "phase-*angle*" ϕ as indicated in this figure, and "phase-plane" as "phase-*space*" in the sense of statistical physics, i.e. a space spanned by coordinates and impulses

we may define the phase ϕ so that the displacement can be written in the form

$$x(t) = A\cos\phi(t)\,. \tag{4.145}$$

The phase ϕ can be easily visualized in two ways, either by the phase shift or by a plot in the so-called phase plane. Let us start with the latter. Here we consider a two-dimensional coordinate system, where the abscissa is the displacement and the ordinate the velocity (Fig. 4.15). In order to obtain a particularly simple situation, instead of the velocity, we rather plot its negative divided by the frequency ω. As one may easily deduce from (4.139) and (4.140), at each time t, $x(t)$ and $-v/\omega$ lie on a circle according to Fig. 4.15. This plot allows us to read off the phase, or in other words, the phase angle ϕ at each time instant. According to (4.144), the phase angle ϕ increases uniformly with velocity ω. The significance of the phase ϕ can also be explained by means of Fig. 4.16 and the concept of a phase shift. Thus instead of (4.139), we consider

$$x(t) = A\cos(\phi(t) + \phi_0) \tag{4.146}$$

which is plotted in Fig. 4.16 and compared with (4.139). The difference between the maxima of both curves is ϕ_0, i.e. the phase-shift. In particular, we find that for

$$\begin{aligned} &\phi_0 > 0, \quad \text{maximum reached at earlier time, and} \\ &\phi_0 < 0, \quad \text{at later time}\,. \end{aligned} \tag{4.147}$$

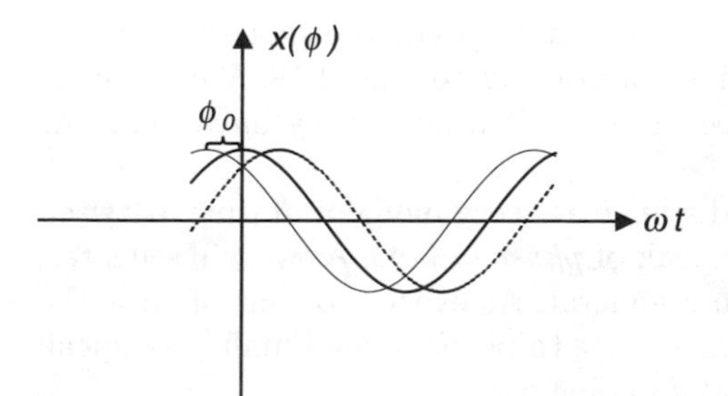

Fig. 4.16. Illustration of phase-shifts: *solid line*: $x = A\cos\omega t$, *dashed line*: $x = A\cos(\omega t - \mid \phi_0 \mid)$; *dotted line*: $x = A\cos(\omega t + \phi_0)$

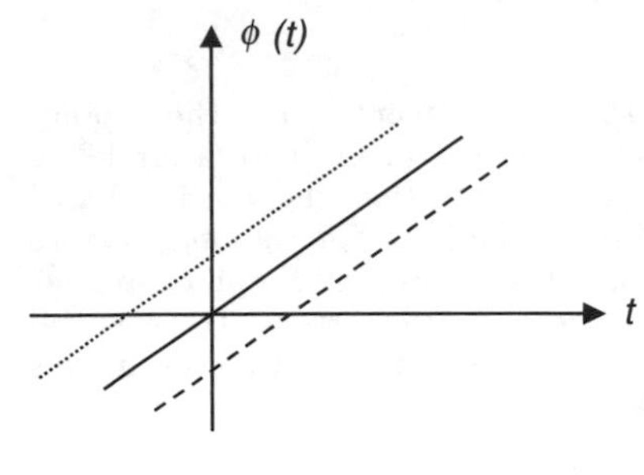

Fig. 4.17. Example of phase- (and frequency-) locking between three oscillators: *solid line*: $\phi = \omega t$, *dashed line*: $\phi = \omega t - \mid \phi_0 \mid$, *dotted line*: $\phi = \omega t + \phi_0$

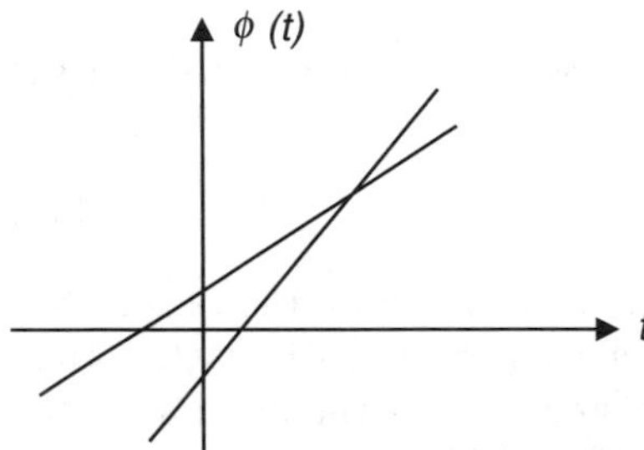

Fig. 4.18. Absence of frequency locking

Taking into account the phase-shift, we may write (4.146) more generally as (4.145), where now

$$\phi(t) = \omega t + \phi_0 . \tag{4.148}$$

holds. The phase velocity is independent of ϕ_0 and reads

$$\dot{\phi}(t) = \omega = \text{const.} \tag{4.149}$$

Plotting $\phi(t)$ versus time t provides us with a set of parallel lines (Fig. 4.17). Such plots become useful when we compare the phases of two or several pendulums or oscillators. In the case of two of them, we have

$$x_1(t) = A_1 \cos \phi_1(t) \tag{4.150}$$

and

$$x_2(t) = A_2 \cos \phi_2(t) . \tag{4.151}$$

Assuming that both ϕs proceed at a constant speed, the plot of $\phi(t)$ versus time t provides us with two lines, according to Fig. 4.18. We speak of *frequency locking* if the two lines become parallel, i.e. if they have the same slope (Fig. 4.19).

If the vertical distance, i.e. the distance at every moment of time, between the two lines remains constant, we speak of *phase locking*. As we will see later, the concept of phase-locking is more general: At every moment of time the distance is the same, but the lines need not to be straight. Finally, we speak of *synchrony* if the two lines ϕ_1 and ϕ_2 coincide.

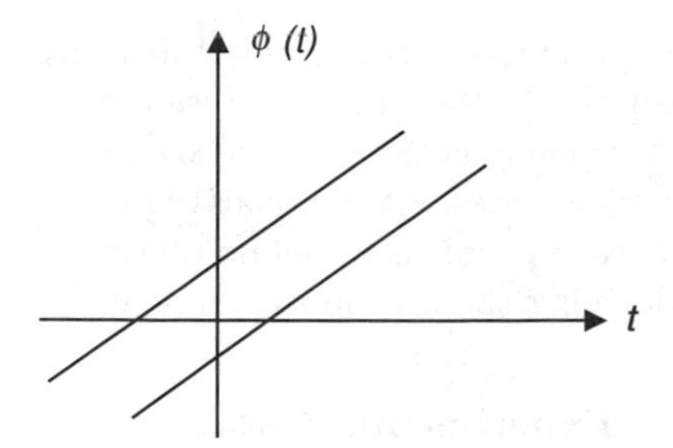

Fig. 4.19. Phase and frequency locking between two oscillators

4.9.2 Regular Spike Trains

While the example of the pendulum gives us a first insight into the significance of the phase, such harmonic oscillations are not the only examples where the concept of phase is highly useful. To illustrate this fact, let us consider regular spike trains, where Fig. 4.20 represents an example. The individual spikes or pulses may have a certain shape, but they may equally well be idealized by δ-functions as we have shown in Sect. 4.1. Now let us consider two spike trains that are plotted in Fig. 4.20. If both spikes occur with the same frequency ω but at different times, we observe a shift ϕ_0 between them, where ϕ_0 quite evidently plays again the role of a phase. If ϕ_0 remains the same over all pairs of spikes stemming from the two individual spike trains, we can again speak of phase-locking. Now let us go a step further, namely let us consider two spike trains, where the sequence of spikes is no more regular but may change either systematically or randomly. Then we may consider two neurons with such spike trains and observe that again the distance between two pairs of such spikes remains constant (Fig. 4.22). Again we may speak of phase-locking in spite of the fact that the frequencies vary.

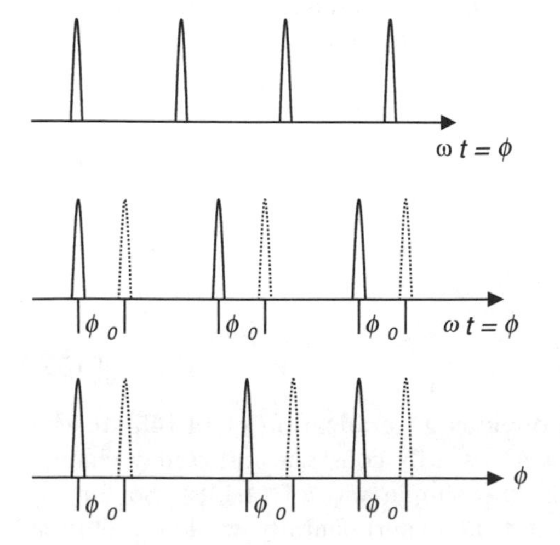

Fig. 4.20. Equidistant pulses seen as a function of phase ϕ

Fig. 4.21. Two equidistant time-shifted pulses seen as a function of phase ϕ

Fig. 4.22. *Solid lines*: An irregular spike train; *dashed line*: The same spike train, but time shifted. While there is no frequency locking, there is phase locking

As we have seen in Chap. 3, phase-locking and more specifically synchrony is an important issue in modern brain research. At the same time we realize that the concept of phase may be useful in harmonic motions, but also equally well in rhythmic motions in which the frequency varies but the relative phases remain constant. Thus it becomes an important problem to define phases in a general way and to devise methods for deriving them from experiments.

4.9.3 How to Determine Phases From Experimental Data? Hilbert Transform

When we measure the axonal pulse of a single neuron, the local field potentials of groups of neurons, or the EEG of still larger groups of them, in each case we have to analyse a certain quantity $x(t)$ that varies in the course of time, and we have to compare such a quantity with a corresponding quantity of another neuron, another group of neurons etc. in order to find out whether phase-locking occurs. There are at least two difficulties: the amplitude A that appears in (4.145) may be time-dependent and the signal x may be corrupted by noise. How can we nevertheless define the phase and furthermore extract it from experimental data? In this endeavour people have resorted to a mathematical tool, namely to the use of the complex plane. This concept is closely related to that of the phase plane used in Sect. 4.9.1. We start with two real variables $x(t)$ and $y(t)$ (for instance displacement and the velocity of a pendulum), use again x as the abscissa, but iy as ordinate, where i is the imaginary unit defined by $i = \sqrt{-1}$. Each point in this *complex plane* can be defined by means of the *complex* variable (see Fig. 4.23)

$$z(t) = x(t) + iy(t) . \tag{4.152}$$

Introducing the distance r and the angle ϕ, we immediately find Fig. 4.24. According to the elementary theory of complex functions, $z(t)$ can be written by means of the distance r and the angle ϕ in the form

$$z(t) = r(t)e^{i\phi(t)} , \tag{4.153}$$

where e is the exponential function. The relation between (4.152) and (4.153) (see also Figs. 4.23 and 4.24) is given by

$$x(t) = r(t) \cos \phi(t) \tag{4.154}$$

and

$$y(t) = r(t) \sin \phi(t) . \tag{4.155}$$

Equation (4.154) represents an obvious generalization of (4.145) to the case where the amplitude $r(t)$ is not necessarily constant and ϕ may change in the course of time other than in the simple way of (4.148). So far, so good; the difficulty rests on the fact that experimentally we know only x

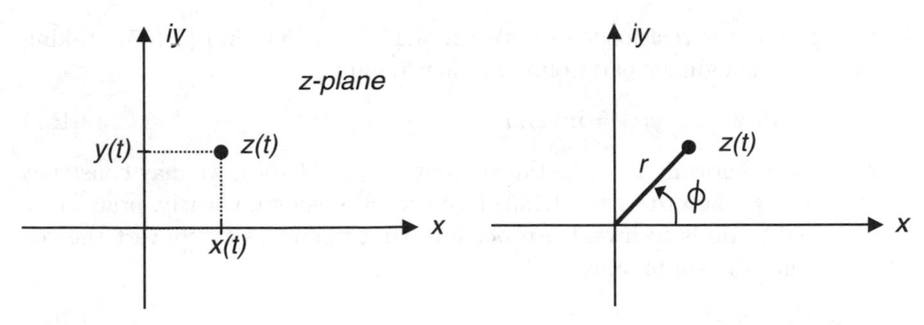

Fig. 4.23. The point $z = x+iy \equiv (x, y)$ in the complex plane

Fig. 4.24. Definition of phase ϕ and radius r in the complex plane of Fig. 4.23

but not y. Is there any way to construct, from the mere knowledge of x, also y, and thus z, and thus the phase angle ϕ? As was found by the famous mathematician Hilbert, this can, indeed, be done. For the experts, we mention that x as measured experimentally must fulfill some specific preconditions, namely in the mathematical sense it must be smooth, i.e. continuously and piecewise differentiable. Let us consider a measurement interval T, which need not be the periodicity interval, however, but must be large enough. Then it is known that we may expand $x(t)$ as a Fourier series

$$x(t) = \sum_{k=0}^{\infty} a_k \cos(\omega kt) + \sum_{k=1}^{\infty} b_k \sin(\omega kt) \tag{4.156}$$

that can also be written in the form

$$x(t) = \sum_{k=0}^{\infty} c_k \cos(\omega kt + \phi_k) \tag{4.157}$$

(see exercises below). Can we derive from this relation one for y,

$$y(t) = \sum_{k=0}^{\infty} c_k \sin(\omega kt + \phi_k) \tag{4.158}$$

by which in each individual term of (4.157) the cos function is replaced by the sin function? If we are able to do so, then we can form

$$z(t) \equiv x(t) + iy(t) = \sum_{k=0}^{\infty} c_k \Big[\cos(\omega kt + \phi_k) + i \sin(\omega kt + \phi_k) \Big] , \tag{4.159}$$

which can be written in the form

$$z(t) = \sum_k c_k e^{i(k\omega t + \phi_k)} \tag{4.160}$$

by using again the relationships between (4.153), (4.154) and (4.155). Taking the real and imaginary parts of z, we then regain

$$x(t) = \mathrm{Re}\, \mathrm{z}(t), \quad y(t) = \mathrm{Im}\, z(t) . \tag{4.161}$$

In particular, knowing z, for instance according to (4.160), we may construct the phase angle according to (4.153) (see exercise below). Clearly, practically all we need to do is to invent a procedure that allows us to convert the cos function into the sin function

$$\cos(\Omega t + \phi) \to \sin(\Omega t + \phi) . \tag{4.162}$$

Since it is sufficient to treat the special case

$$\cos t \to \sin t \tag{4.163}$$

we treat that case here and refer the reader with respect to the general case (4.162) to the exercises.

According to Hilbert, we consider

$$P \int_{-\infty}^{+\infty} \frac{\cos \tau}{\tau - t} d\tau , \tag{4.164}$$

where the P in front of the integral refers to the principal value. Clearly, for $\tau = t$, the integrand diverges. According to the principal value, we cut that singularity out by defining

$$P \int_{-\infty}^{+\infty} ..d\tau = \int_{-\infty}^{-\epsilon} ..d\tau + \int_{\epsilon}^{\infty} ..d\tau , \tag{4.165}$$

i.e. we cut out the immediate surrounding of the singularity taking finally the limit $\epsilon \to 0$. In (4.164) we introduce a new variable by means of

$$\tau - t = \sigma \tag{4.166}$$

so that we obtain

$$P \int_{-\infty}^{+\infty} \frac{\cos(\sigma + t)}{\sigma} d\sigma . \tag{4.167}$$

In it we use an elementary formula from trigonometry, namely

$$\cos(\sigma + t) = \cos \sigma \cos t - \sin \sigma \sin t \tag{4.168}$$

so that we can cast (4.167) into the form

$$P \int_{-\infty}^{+\infty} \frac{\cos \sigma}{\sigma} d\sigma \cos t - P \int_{-\infty}^{+\infty} \frac{\sin \sigma}{\sigma} d\sigma \sin t . \tag{4.169}$$

Because of symmetry (see exercises), the first integral vanishes. The integrand of the integral in

$$P \int_{-\infty}^{+\infty} \frac{\sin \sigma}{\sigma} d\sigma = \int_{-\infty}^{+\infty} \frac{\sin \sigma}{\sigma} d\sigma = 2\pi \tag{4.170}$$

remains finite at $\sigma = 0$. This integral can be found in mathematical tables and is given according to (4.170). Putting the individual steps (4.165)–(4.170) together, we obtain the decisive result

$$P \int_{-\infty}^{+\infty} \frac{\cos \tau}{\tau - t} d\tau = -2\pi \sin t \, . \tag{4.171}$$

This relation provides us with the "recipe" to transform a cos function into a sin function! Taking care of the change of sign and of the factor 2π, we obtain for each individual member of the series (4.157) and (4.158), and thus for the whole series, also the relationship

$$y(t) = \frac{1}{2\pi} P \int_{-\infty}^{+\infty} \frac{x(\tau)}{t - \tau} d\tau \, . \tag{4.172}$$

This is the fundamental Hilbert transform that allows us to derive the additional function y if the function x is given. Making use of (4.154) and (4.155), we may now determine the phase ϕ by means of

$$\sin \phi / \cos \phi \equiv \mathrm{tg}\phi = y/x \, , \tag{4.173}$$

or explicitly by

$$\phi = \mathrm{arc} \tan(y/x) \, . \tag{4.174}$$

Plotting ϕ versus t, for instance for two series of experimental data (see Fig. 4.6), we may discuss phase locking, frequency locking and synchrony in generalization of what we have discussed in Sect. 4.9.1. At least two difficulties should be mentioned here when applying this procedure to realistic experimental data:

1) The observation time T is only finite, whereas the Hilbert transform (4.172) requires an infinite interval because of the integration in that formula. Thus in concrete cases one has to study how well this formula works in the case of a finite observation interval.
2) In general, the experimental data are noisy, which implies that in many cases $x(t)$ is not smooth. Therefore, it is important to use smoothing procedures, which we will not discuss here, however.

Exercise.

1. Show that (4.156) can be written as (4.157).
   Hint: Use $\cos(\alpha+\beta) = \cos\alpha\cos\beta - \sin\alpha\sin\beta$.
2. Extend the procedure related to (4.163) to one for (4.162).
   Hint: Introduce a new variable $t' = \omega t$.
3. Show that

   $$P\int_{-\infty}^{+\infty} \frac{\cos\sigma}{\sigma} d\sigma$$

   vanishes because of symmetry.
   Hint: Make the transformation $\sigma \to -\sigma$.
4. Convince yourself of the correctness of (4.172) using Exercise 2.
5. Discuss frequency locking, phase locking and synchrony by means of Fig. 4.6.
   Hint: Plot $\xi(t) = \phi_2 - \phi_1$ versus t.

4.10 Phase Noise

As we have seen earlier in this chapter, noise in any physical system including the brain is inevitable. This implies also that phases are noisy. In this section, we want to show how we can model the origin and important properties of phase noise. In the most simple case, we may assume that by analogy with the soccer game the phase is subjected to random kicks. As we will see later in this section, the phase need not be damped so that the change of the phase in the course of time is merely determined by a random force $F(t)$ alone

$$\dot{\phi}(t) = F(t) . \tag{4.175}$$

As usual we assume that the average over F vanishes

$$< F(t) >= 0 \tag{4.176}$$

and that the correlation function is δ-correlated with strength Q

$$< F(t)F(t') >= Q\delta(t-t') . \tag{4.177}$$

Equation (4.175) can be immediately solved by integration over time on both sides, yielding

$$\phi(t) - \phi(t_0) = \int_{t_0}^{t} F(\tau)d\tau . \tag{4.178}$$

Taking the statistical average over both sides of (4.178) and using (4.176), we obtain

$$< \phi(t) > -\phi(t_0) = 0 , \tag{4.179}$$

i.e. on the average the phase remains unchanged. In order to define a measure telling us how far the phase deviates in the course of time, we form the quadratic displacement, which can easily be evaluated according to

$$< (\phi(t) - \phi(t_0))^2 > = < \int_{t_0}^{t} F(\tau)d\tau \int_{t_0}^{t} F(\tau')d\tau' > , \tag{4.180}$$

$$= \int_{t_0}^{t}\int_{t_0}^{t} < F(\tau)F(\tau') > d\tau d\tau' , \tag{4.181}$$

$$= \int_{t_0}^{t}\int_{t_0}^{t} Q\delta(\tau - \tau')d\tau d\tau' = Q \int_{t_0}^{t} d\tau = Q(t - t_0) , \tag{4.182}$$

where the individual steps are entirely analogous to what we have done in Sect. 4.7 on Brownian motion. The result (4.182) is interpreted as phase diffusion, where the mean square of the displacement of the phase linearly increases in the course of time.

Let us consider the next simple case that is defined by

$$\dot{\phi}(t) = \omega + F(t) , \tag{4.183}$$

where, quite evidently, on the average the phase velocity is constant. Integration of (4.183) on both sides yields

$$\phi(t) - \phi(t_0) = \omega(t - t_0) + \int_{t_0}^{t} F(\tau)d\tau , \tag{4.184}$$

from which we immediately deduce

$$< \phi(t) - \phi(t_0) > = \omega(t - t_0) \tag{4.185}$$

as well as

$$< (\phi(t) - \phi(t_0) - \omega(t - t_0))^2 > = Q(t - t_0) . \tag{4.186}$$

Equation (4.186) can be rearranged to yield

$$< (\phi(t) - \phi(t_0)^2 > = \omega^2(t - t_0)^2 + Q(t - t_0) . \tag{4.187}$$

Thus the mean quadratic displacement of ϕ in the course of time is determined by two effects, namely the quadratic increase according to the deterministic motion and a linear increase according to stochastic pushes.

As we will see later, in practical applications also a third case appears that is defined by

$$\dot{\phi} = \omega - \gamma \sin\phi + F(t) . \tag{4.188}$$

In applications quite often ϕ is the relative phase between two oscillators. To study (4.188), we first assume that the random force F vanishes

$$F(t) = 0 \tag{4.189}$$

and that there is a time-independent solution

$$\dot{\phi} = 0 \tag{4.190}$$

so that we have to solve

$$\omega - \gamma \sin \phi_0 = 0 \,. \tag{4.191}$$

In fact there exists a solution provided

$$\omega < \gamma \tag{4.192}$$

holds. The solution is given by

$$\phi_0 = \text{arc } \sin(\omega/\gamma) \tag{4.193}$$

and if ϕ is a relative phase, phase locking occurs. In order to study the impact of the fluctuating force on this phase-locked solution, we make the hypothesis

$$\phi(t) = \phi_0 + \psi(t) \tag{4.194}$$

and assume that F is small so that also the deviation ψ will become small. Using the well-known trigonometric formula

$$\sin \phi = \sin(\phi_0 + \psi) = \sin \phi_0 \cos \psi + \cos \phi_0 \sin \psi \tag{4.195}$$

and approximating $\cos \psi$ and $\sin \psi$ up to linear terms in ψ, (4.195) reduces to

$$\sin(\phi_0 + \psi) \approx \sin \phi_0 + \psi \cos \phi_0 \,. \tag{4.196}$$

Inserting (4.194) into (4.188) and making use of (4.191), we readily obtain

$$\dot{\psi} = \omega - \gamma \sin \phi_0 - \psi \cos \phi_0 + F(t) \,. \tag{4.197}$$

Using the abbreviation

$$\cos \phi_0 = \gamma' \,, \tag{4.198}$$

(4.197) can be cast into the form

$$\dot{\psi} = -\gamma' \psi + F(t) \,. \tag{4.199}$$

This equation is an old acquaintance of ours provided we identify ψ with the displacement of a particle x

$$\psi(t) \leftrightarrow x(t) \tag{4.200}$$

that we encountered in the section on Brownian motion. Thus the phase deviation ψ undergoes a Brownian motion in the case of phase locking. So far we have assumed (4.192). If this condition is not fulfilled, i.e. if γ is small and ω large, we are practically back to the case (4.183), and we will not discuss this case here in more detail though there are some slight modifications because of the additional effects of the small sin function in (4.188).

4.11 Origin of Phase Noise*

Later in this book we will be concerned with concrete models on the activity of individual neurons or networks of them. In other words, we will study the dynamics of such systems. A famous model that deals with the generation of axonal pulses is provided by the Hodgkin–Huxley equations (see Chap. 11). In order not to overload our presentation here, we mention only a few facts. These equations describe how the membrane voltage changes in the course of time because of in- or outflow of ions, for instance calcium ions. On the other hand, they describe how the calcium flux changes in the course of time, because of externally applied currents as well as of the voltage change. Thus, in the simplest case, we have to deal with two variables x (membrane voltage) and y (ion fluxes). But by means of two variables x and y we may form the complex quantity z, and thus according to

$$z(t) = r(t)e^{i\phi(t)} \tag{4.201}$$

determine a phase. Quite surprisingly, we can derive a few general properties of the equation for ϕ without knowing details of, say, the Hodgkin–Huxley equations. We need to know that a membrane voltage as well as the ion currents change in the course of time, because of their present values. Thus, quite generally, we have to deal with equations of the form

$$\dot{x} = g(x, y) + F_x \tag{4.202}$$

and

$$\dot{y} = h(x, y) + F_y \, , \tag{4.203}$$

where the first term on the r.h.s. describes the noiseless case. F_x and F_y are the fluctuating forces, whose origin we discussed in previous sections on Brownian motion and on noise in active systems. By multiplying (4.203) with the imaginary unit i and adding this equation to (4.202), we obtain

$$\dot{z} + H(z, z^*) = F(t) \, , \tag{4.204}$$

where the complex variable z is defined as usual (see (4.152)). We use the abbreviations

$$H = -g - ih \tag{4.205}$$

and

$$F = F_x + iF_y \tag{4.206}$$

with the property

$$< F(t)F(t') >= Q\delta(t-t') . \tag{4.207}$$

For what follows, we decompose H into a linear and a nonlinear part

$$H(z,z^*) = cz + dz^* + N(z,z^*) , \tag{4.208}$$

where

$$c = \gamma - i\omega . \tag{4.209}$$

Inserting (4.201) into (4.204) and using (4.208), we obtain after multiplying the resulting equation by $e^{-i\phi}$

$$\dot{r} + ri\dot{\phi} + cr + de^{-2i\phi} + e^{-i\phi}N\left(re^{i\phi}, re^{-i\phi}\right) = e^{-i\phi}F(t) . \tag{4.210}$$

The l.h.s. may be expanded into positive and negative powers of $e^{i\phi}$, which yields

$$\dot{r} + ri\dot{\phi} + cr + h(r) + g_1(r)e^{i\phi} + g_2(r)e^{-i\phi} + .. = e^{-i\phi}F(t) . \tag{4.211}$$

In order not to overload our presentation, we consider a special case, namely in which the coefficients of the powers of $e^{i\phi}$ and $e^{-i\phi}$ are small. In such a case, (4.211) reduces to

$$\dot{r} + ir\dot{\phi} + (\gamma - i\omega)r + h(r) = e^{-i\phi}F(t) . \tag{4.212}$$

Splitting (4.212) into its real and its imaginary parts, we obtain

$$\dot{r} + \gamma r + h(r) = \mathrm{Re}\, F(t) \tag{4.213}$$

and

$$\dot{\phi} = \omega + \frac{1}{ir}\mathrm{Im}\, F(t) , \tag{4.214}$$

where

$$\mathrm{Re}\, F = \frac{1}{2}(F + F^*) , \quad \mathrm{Im}\, F = \frac{1}{2}(F - F^*) . \tag{4.215}$$

Equation (4.214) represents the desired formula for the impact of noise on the phase movement. If we were to keep powers of exp $(i\phi)$ in (4.211), cos and sin functions would appear on the r.h.s. of (4.214) also. Because of the function r that appears in (4.214), the noise force depends on r. In a number of cases, r can be considered as practically constant so that (4.214) reduces to the former case (4.183).

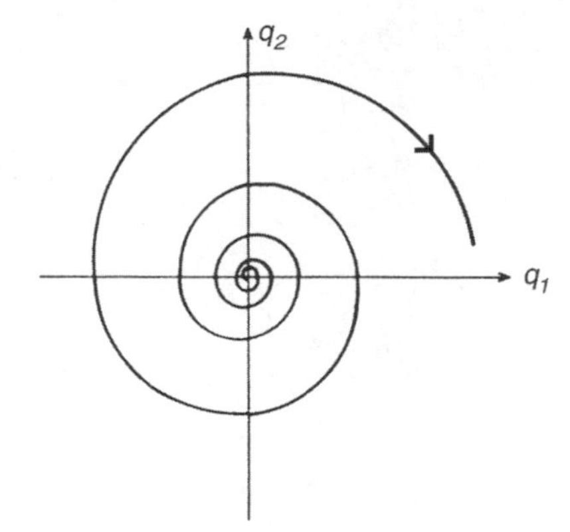

Fig. 4.25. A trajectory starting from an unstable "focus" and approaching a stable "limit cycle"

For the sake of completeness, we present an example where r is time-dependent, but relaxes towards a constant value. It must be stated, however, that this case does not apply to the Hodgkin–Huxley equations, where we rather have to expect a strong time-dependence of r because of the spikes! Nevertheless the reader may enjoy this example.

We consider the noiseless case of (4.213), where the r.h.s. vanishes. The simplest nontrivial example of this equation is provided by

$$\dot{r} + (r - a)r = 0, \quad a > 0\,, \tag{4.216}$$

that possesses the solution

$$r = \frac{a r_0}{r_0 + (a - r_0)\exp[a(t_0 - t)]}\,, \tag{4.217}$$

where r_0 represents the initial value

$$r(t_0) = r_0 > 0\,. \tag{4.218}$$

As one may easily verify, in the limit $t \to \infty$, $r(t)$ relaxes towards a, i.e. to a constant value. Jointly with the movement of the phase this represents what is called a *stable limit cycle* (Fig. 4.25). Readers who enjoy mathematical exercises can derive (4.217) from (4.216) (see exercise).

Exercise. Derive the solution (4.217) to (4.216).

Part II

Spiking in Neural Nets

5. The Lighthouse Model. Two Coupled Neurons

5.1 Formulation of the Model

In this chapter the main part of our book begins. It will be our goal to develop models that allow us to study the behavior of large neural nets *explicitly*. Hereby we use essential properties of neurons. Both this and the following chapter deal with what I call the lighthouse model for reasons that I will explain below.

To model the behavior of a *single neuron*, we start from a few basic facts (see Chap. 2). A neuron receives inputs from other neurons in the form of their axonal spikes. At synapses these spikes are converted into dendritic currents that lead to a potential change at the axon hillock. Such a conversion of spikes into potential changes can also happen directly and will be contained in our formalism as a special case. In response to incoming signals, the neuron produces an output in the form of axonal pulses (spikes).

Let us begin with the conversion of a spike into a dendritic current (Fig. 5.1). We label the dendrite under consideration by an index, m, and denote the corresponding current at time t by $\psi_m(t)$ (Fig. 5.2). The current is caused by a spike that was generated at the axon hillock of another neuron, labelled by k, at a somewhat earlier time, τ. The spike or pulse train is represented by a function $P_k(t-\tau)$ (Fig. 5.3). After its generation, the dendritic current will decay with a decay constant γ (Fig. 5.4). We shall assume that γ is independent of m. As we know, all physical systems are subject to noise, whose effect we take into account by a fluctuating force that we call $F_{\psi,m}(t)$.

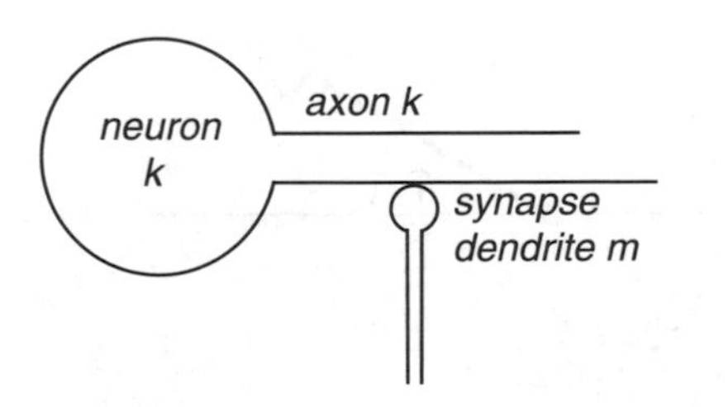

Fig. 5.1. Neuron k with its axon k connected to a dendrite m of a different neuron via a synapse

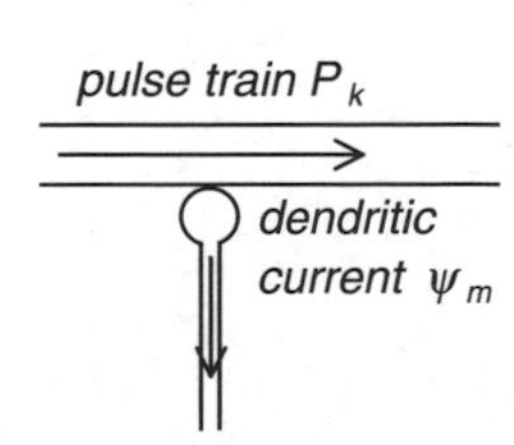

Fig. 5.2. A pulse train P_k in axon k causes a dendritic current ψ_m via a synapse

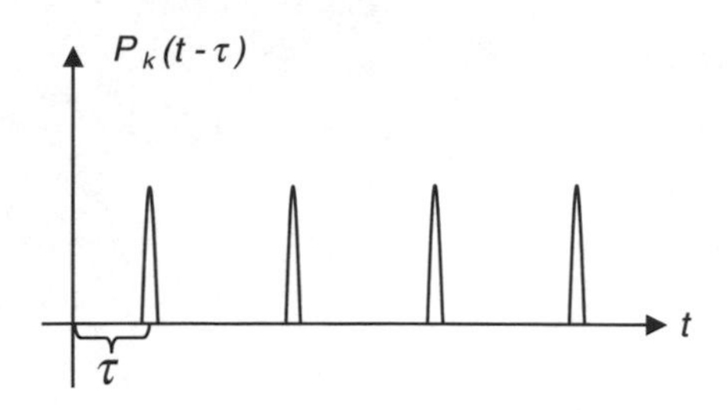

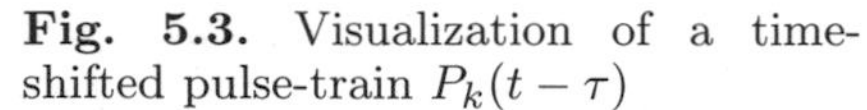

Fig. 5.3. Visualization of a time-shifted pulse-train $P_k(t-\tau)$

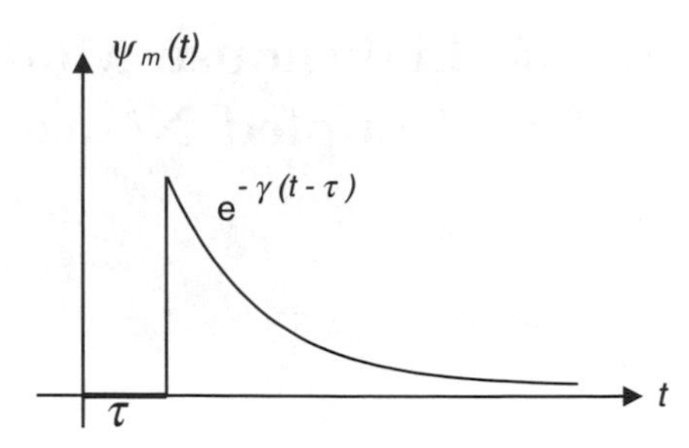

Fig. 5.4. The dendritic current $\psi_m(t)$

We discussed such forces in Sects. 4.6–4.8. Readers who haven't read these sections or are not interested in these forces at all can ignore them in most parts of what follows. Using the results of Sect. 4.4 we are now in a position to formulate the corresponding equation for the dendritic current ψ as follows:

$$\dot{\psi}_m(t) = aP_k(t-\tau) - \gamma\psi_m(t) + F_{\psi m}(t) \, . \tag{5.1}$$

The constant a represents the synaptic strength. For the interested reader we add a few more comments on F_ψ. As usual we shall assume that the fluctuating forces are δ-correlated in time. As is known, in the synapses vesicles that release neurotransmitters and thus eventually give rise to the dendritic current can spontaneously open. This will be the main reason for the fluctuating force F_ψ. But also other noise sources may be considered here. When a pulse comes in, the opening of vesicles occurs with only some probability. Thus we have to admit that in a more appropriate description a is a randomly fluctuating quantity. While $F_{\psi m}$ in (5.1) represents additive noise, a represents multiplicative noise.

Since one of our main tasks is the study of synchronization between neurons, it suggests itself to introduce a phase angle by which we describe spike trains. The basic idea is this: Imagine a lighthouse with its rotating light beam (Fig. 5.5). Whenever the beam hits a fixed target, we observe a flash (Fig. 5.6). Depending on the beam's rotation speed, we obtain more or less rapid sequences of flashes. Using this picture, we introduce – formally

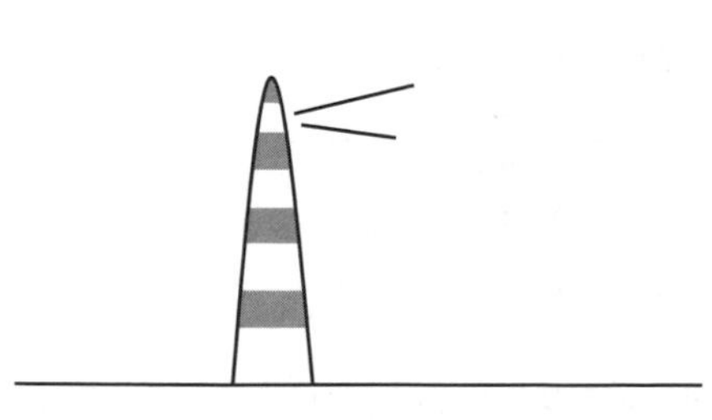

Fig. 5.5. Lighthouse with rotating light beam

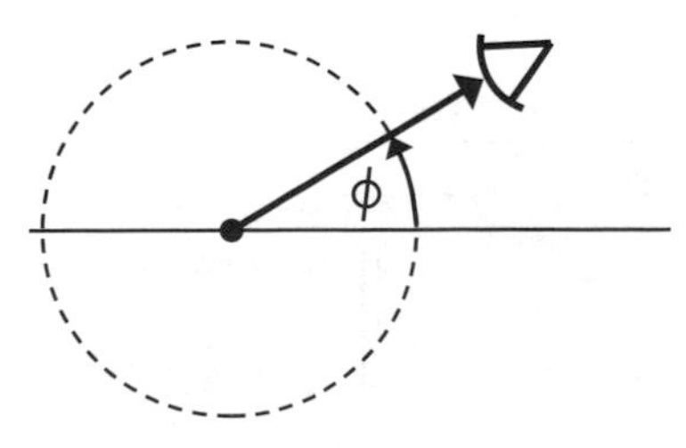

Fig. 5.6. The light beam of Fig. 5.5, seen from above, hits a target at a specific angle ϕ

– a phase ϕ of the neuron under consideration. This phase increases in the course of time and whenever it equals 2π or an integer multiple of 2π, an axonal pulse or spike is generated. In this way a spike train is brought about. Each individual spike is modelled by means of a δ-function according to Sect. 4.1. Hereby we have two choices. Either we consider δ as a function of time t with its peak at

$$t = t_n\,, \quad t_n \quad \text{fixed}\,, \tag{5.2}$$

or as a function of $\phi(t)$ with its peak at

$$\phi(t_n) = 2\pi n\,. \tag{5.3}$$

In this latter case, the handling of $\delta(\phi(t) - 2\pi)$ becomes more involved. We therefore use a "peaked function" in accordance with Sect. 4.1

$$\delta(\phi(t) - 2\pi n)\dot{\phi}(t)\,. \tag{5.4}$$

Both (5.4) and $\delta(t - t_n)$ possess a single peak with "area" 1, i.e.

$$\int \delta(t - t_n)dt = \int \delta(\phi(t) - 2\pi n)\dot{\phi}(t)dt = 1\,. \tag{5.5}$$

In order to represent all spikes in the course of time, we sum up over all peaks, i.e. we form

$$P(t) \equiv f(\phi(t)) = \sum_n \delta(\phi(t) - 2\pi n)\dot{\phi}(t_n)\,, \tag{5.6}$$

where n runs over all integers, in principle from $-\infty$ to $+\infty$. When P and the phase ϕ refer to a neuron with index k, we have to supplement P and ϕ with the index k. So far, the introduction of the phase ϕ is just a formal trick to describe spike-trains. The essential question is, of course, how can we determine the time-dependence of ϕ? Here the Naka–Rushton formula (Sect. 2.5) comes in. According to it, the axonal spike rate S is determined by the input X to the neuron, i.e. according to (2.1) by

$$S(X) = \frac{rX^M}{\Theta^M + X^M}\,. \tag{5.7}$$

But, at least under steady-state conditions, the spike rate is directly proportional to the "rotation speed" $\dot{\phi}$. Absorbing the proportionality factor 2π into a redefined constant r in (5.7), we may thus put

$$\dot{\phi} = S(X)\,. \tag{5.8}$$

Because (5.8) refers to a physical system, the neuron, fluctuating forces must be taken into account. Furthermore, we must add the label k so that the equation for the phase of neuron k reads

$$\dot{\phi}_k = S(X_k) + F_{\psi,k}(t)\,. \tag{5.9}$$

It remains to determine the input X_k that leads to the spike generation. As we know, spikes are generated at the axon hillock at which the potentials due to the dendritic currents are added. We may take time delays τ' into account as well as coefficients c that convert ψ into potential contributions. Also external signals, $p_{\text{ext}}(t)$, stemming from sensory neurons must be taken into account. Thus, all in all, we arrive at

$$X_k(t) = \sum_m c_{km}\psi_m(t-\tau') + p_{\text{ext,k}}(t-\tau'') . \tag{5.10}$$

This concludes the formulation of our model in its most simple form.

To summarize, our model is defined by (5.1), (5.6), (5.7), (5.9) and (5.10). In order to familiarize ourselves with it, we shall consider the case of two neurons in the present chapter. In Chap. 6 we shall treat the general case with an arbitrary number of neurons and dendrites, with different synaptic strengths, and different delay times. Even in this rather general model, a few limitations must be observed. The use of the Naka–Rushton relation (or similar ones) implies steady states. Otherwise, the validity of our approach must be checked in individual cases. More seriously according to (5.1), the dendritic current is spontaneously generated (see Sect. 4.4). In reality, the current first increases continuously until it starts its decay. This is taken care of by Rall's α-function that we mentioned in Sect. 4.4. Finally, we may try to explore the physical meaning of ϕ and – as a consequence – to introduce a damping term (5.9). These deficiencies will be remedied in Chap. 8, but the treatment of the corresponding equations will become cumbersome and we shall have to restrict ourselves to the most important aspects.

But now let us return to the lighthouse model for two neurons.

5.2 Basic Equations for the Phases of Two Coupled Neurons

We consider two neurons, $k = 1, 2$, each with one dendrite, $m = 1, 2$, that are mutually coupled (Fig. 5.7). We assume that the system operates in the linear regime of S, which according to Sect. 2.5 is quite a good approximation for inputs that are not too high. We neglect delays, i.e. we put $\tau = \tau' = 0$ and ignore fluctuating forces, i.e. we put $F_\psi = F_\phi = 0$. The model equations introduced in the foregoing section become very simple!

neuron 1:

$$\text{dendrite}: \quad \dot{\psi}_1(t) = af(\phi_2(t)) - \gamma\psi_1 ; \tag{5.11}$$

$$\text{axon}: \quad \dot{\phi}_1(t) = c\psi_1(t) + p_{\text{ext},1} . \tag{5.12}$$

neuron 2:

$$\text{dendrite}: \quad \dot{\psi}_2 = af(\phi_1(t)) - \gamma\psi_2 ; \tag{5.13}$$

$$\text{axon}: \quad \dot{\phi}_2 = c\psi_2(t) + p_{\text{ext},2} . \tag{5.14}$$

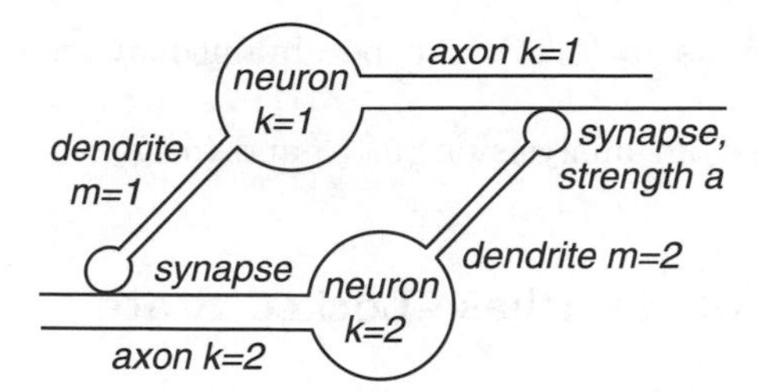

Fig. 5.7. Scheme of two coupled neurons

Using (5.12), we express ψ_1 by ϕ_1 and insert the result into (5.11), thus obtaining

$$\ddot{\phi}_1 + \gamma\dot{\phi}_1 = Af(\phi_2) + C_1 \tag{5.15}$$

for neuron 1, where

$$A = ca\,, \quad C_1 = \gamma p_{\text{ext},1} + \dot{p}_{\text{ext},1}\,. \tag{5.16}$$

Similarly, we obtain for neuron 2

$$\ddot{\phi}_2 + \gamma\dot{\phi}_2 = Af(\phi_1) + C_2\,, \tag{5.17}$$

where

$$C_2 = \gamma p_{\text{ext},2} + \dot{p}_{\text{ext},2}\,. \tag{5.18}$$

There is yet another, though equivalent, way of eliminating the dendritic currents. Namely, the Green's function method of Sect. 4.5 (see (4.83), (4.95) and (4.96)) allows us to express the dendritic current ψ_1 in (5.11) by the phase ϕ_2, i.e. the formal solution to (5.11) reads

$$\psi_1(t) = a\int_0^t e^{-\gamma(t-\sigma)} f(\phi_2(\sigma))d\sigma\,. \tag{5.19}$$

Inserting this into (5.12) yields

$$\dot{\phi}_1(t) = ac\int_0^t e^{-\gamma(t-\sigma)} f(\phi_2(\sigma))d\sigma + p_{\text{ext},1}(t)\,. \tag{5.20}$$

This equation allows for a particularly simple interpretation: The rotation speed $\dot{\phi}_1$ is caused by the sum of the input from the other neuron and the external signal. If we express $f(\phi_2)$ by the emission times $t_{2,n}$ of the pulses stemming from neuron 2, f can be replaced by a sum of δ-functions (see Sect. 4.1)

$$\dot{\phi}_1(t) = ac\int_0^t e^{-\gamma(t-\sigma)} \sum_n \delta\left(\sigma - t_{2,n}\right) d\sigma + p_{\text{ext},1}(t)\,. \tag{5.21}$$

The equation for ϕ_2, which corresponds to (5.20), can be obtained by exchanging the indices 1 and 2.

In the following, we shall first base our analysis on (5.15) and (5.17).

5.3 Two Neurons: Solution of the Phase-Locked State

In this section we wish to derive the solution to (5.15) and (5.17) which belongs to the phase-locked state. Since we do not yet know under which conditions, i.e. for which parameter values of γ, A and C_1, C_2, this state exists, we consider a case in which we surely can expect phase locking: When both neurons are subject to the same conditions. This is the case if $C_1 = C_2 = C$. Indeed, such a choice is suggested by the experimental conditions, when the same moving bar is shown to the receptive fields to the two neurons. Phase locking, and even more, synchrony occurs, if the phases ϕ_1 and ϕ_2 coincide, i.e.

$$\phi_1 = \phi_2 = \phi \,. \tag{5.22}$$

Making this substitution in (5.15) and (5.17), the two equations acquire precisely the same form, namely

$$\ddot{\phi} + \gamma\dot{\phi} = Af(\phi) + C \,. \tag{5.23}$$

Because the function $f(\phi)$ on the r.h.s. of (5.23) depends on ϕ in a highly nonlinear fashion, the explicit solution of (5.23) might look hopeless.

So let's first try to study what it will look like by means of an analogy. To a physicist, (5.23) may be reminiscent of an equation in mechanics, to an electrical engineer of that of an electrical circuit, and so on. Let's adopt the physicist's view and identify the phase ϕ with the position coordinate x of a particle. The equation coming to his/her mind reads

$$m\ddot{x} + \gamma\dot{x} = F(x) \,, \tag{5.24}$$

where m is the mass of the particle, γ the constant of friction and $F(x)$ a force acting on the particle. In the present case we put $m = 1$ and identify $F(x)$ with $Af(x) + C$. But a physicist will go a step further. He/she remembers that the force $F(x)$ is nothing but the slope of a (mechanical) potential $V(x)$ multiplied by -1:

$$F(x) = -\frac{\partial V(x)}{\partial x} \,. \tag{5.25}$$

Let us first consider an example where the force F is independent of x

$$F(x) = C \,. \tag{5.26}$$

The potential V is plotted in Fig. 5.8. The movement of the particle, i.e. the change of its position x in the course of time, can be understood as

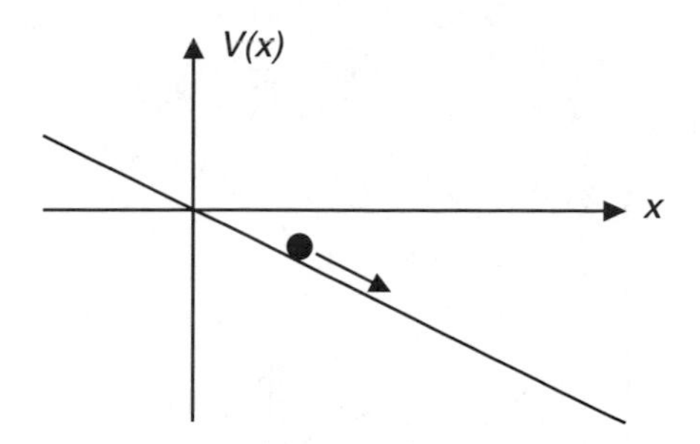

Fig. 5.8. Ball sliding down the potential curve $V(x) = -Cx$

follows: The particle behaves like a ball that rolls down the potential "hill" being subject to friction. First the velocity will also increase until the friction force $\gamma\dot{x}$ compensates the constant force $F = C$ so that

$$\gamma\dot{x} = C \quad \text{or} \quad \dot{x} = C/\gamma\,. \tag{5.27}$$

Integrating (5.27) yields

$$x(t) = (C/\gamma)t\,. \tag{5.28}$$

Equations (5.27) and (5.28) tell us that our ball is rolling down the potential "hill" at constant speed.

Let us now return to our horrible-looking equation (5.23). So what is the effect of

$$f(\phi) = \dot{\phi}\sum_n \delta(\phi - \phi_n) \tag{5.29}$$

where we return to our old notation of ϕ instead of x? Because f does not only depend on ϕ, but also on the velocity $\dot{\phi}$, we cannot directly use the potential $V(\phi)$ by analogy with (5.25). But we may, at least tentatively, assume that in view of the effect of damping (see (5.27)), $\dot{\phi}$ is, at least on the average, constant. The resulting function $f(\phi)$ consists of a sum over δ-functions, i.e. of a sum over individual peaks. The corresponding potential function is shown in Fig. 5.9. Roughly speaking, it looks similar to that of Fig. 5.8 except for individual jumps. Each time ϕ (or formerly x) reaches such a position, the "force" f gives a push to the phase (particle). Afterwards, because of friction, it will slow down again towards its previous velocity until it will get a new push, a.s.o.

We now want to show how our intuitive view can be cast into a rigorous mathematical approach. To take the overall constant motion into account, we put

$$\phi = \frac{1}{\gamma}Ct + \chi = ct + \chi\,, \tag{5.30}$$

where χ is a still unknown function of time. Inserting (5.30) into (5.23) yields

$$\ddot{\chi} + \gamma\dot{\chi} = Af(\chi + ct)\,. \tag{5.31}$$

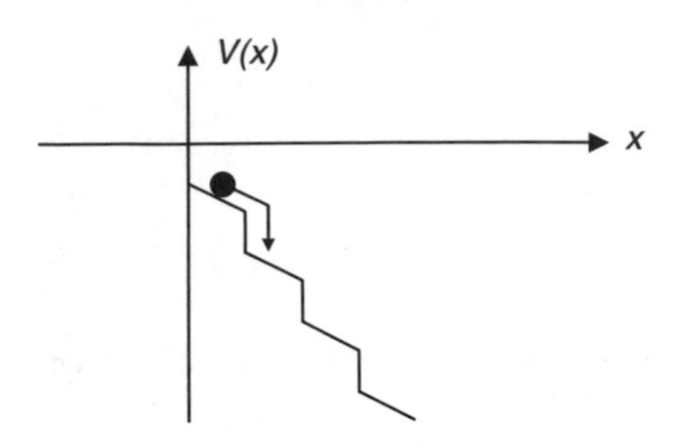

Fig. 5.9. Ball sliding (and jumping) down the potential curve corresponding to (5.29) with $\dot{\phi} = \text{const.}$

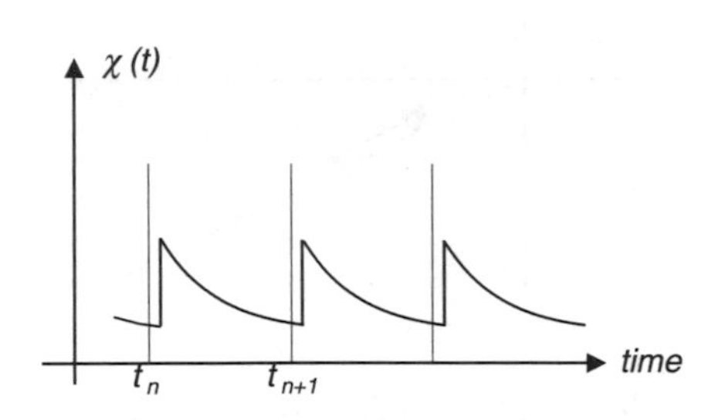

Fig. 5.10. The function $\chi(t)$

We expect that when the phase ϕ grows in time, it will receive pushes at the peaks of the δ-functions, which happen at phase positions $\phi = \phi_n$ or, equivalently, times $t = t_n$ (Fig. 5.10). In between these times, the force in (5.31) vanishes. Thus for the time interval $t_n + \epsilon \leq t \leq t_{n+1} - \epsilon$, where ϵ is arbitrarily small, we obtain

$$\ddot{\chi} + \gamma\dot{\chi} = 0\,, \tag{5.32}$$

which has the solution

$$\dot{\chi}(t) = \dot{\chi}\,(t_n + \epsilon) \cdot e^{-\gamma(t-t_n)}\,. \tag{5.33}$$

We are left with studying what happens at times $t = t_n$. We note that due to a general mathematical result, $\chi(t)$ is continuous for all t, i.e.

$$\chi(t + \epsilon) = \chi(t - \epsilon)\,, \quad \epsilon \to 0\,. \tag{5.34}$$

To cope with the δ-functions occurring in $f(\phi)$, we integrate (5.31) over a small time interval

$$(t_n - \epsilon, t_n + \epsilon) \tag{5.35}$$

and obtain

$$\int\limits_{t_n-\epsilon}^{t_n+\epsilon} \ddot{\chi}(t)dt + \gamma \int\limits_{t_n-\epsilon}^{t_n+\epsilon} \dot{\chi}(t)dt = A \int\limits_{t_n-\epsilon}^{t_n+\epsilon} f(\phi(t))dt\,. \tag{5.36}$$

Because of the derivatives under the integrals on the l.h.s. of (5.36), we can perform the integration immediately and obtain

$$\dot{\chi}(t_n + \epsilon) - \dot{\chi}(t_n - \epsilon) + \gamma(\chi(t_n + \epsilon) - \chi(t_n - \epsilon))\,, \tag{5.37}$$

where the factor of γ vanishes because of (5.34). On the r.h.s. of (5.36), because of the "peaky" character of f, we need to consider only a single term,

$$A \int_{t_n-\epsilon}^{t_n+\epsilon} \dot{\phi}(t)\delta(\phi(t)-\phi_n)dt\,. \tag{5.38}$$

But as we have seen in Sect. 4.4, the expression under the integral is constructed in such a way that the integral yields one. Thus the relation (5.37), eventually, reduces to

$$\dot{\chi}(t_n+\epsilon)-\dot{\chi}(t_n-\epsilon)=A\,. \tag{5.39}$$

Using (5.33) in (5.39), we obtain the recursive relation

$$\dot{\chi}\,(t_{n+1}+\epsilon)=\dot{\chi}\,(t_n+\epsilon)\,e^{-\gamma(t_{n+1}-t_n)}+A\,. \tag{5.40}$$

We first assume that the times t_n are given quantities. While we postpone the general solution of (5.40) to the exercise at the end of the section, we focus our attention on the steady-state solution of that equation. In this case, the times t_n are equidistant,

$$t_{n+1}-t_n=\Delta\,,\quad \text{independent of}\quad n\,, \tag{5.41}$$

and

$$\dot{\chi}(t_{n+1}+\epsilon)=\dot{\chi}(t_n+\epsilon)=\dot{\chi}\,. \tag{5.42}$$

Inserting (5.42) into (5.40) leads to the solution of (5.40),

$$\dot{\chi}=A\left(1-e^{-\gamma\Delta}\right)^{-1}\,. \tag{5.43}$$

So far we assumed the "jump times" t_n to be given. In order to determine them, we remember that from spike to spike, or from push to push, the phase increases by 2π, which means

$$\int_{t_n+\epsilon}^{t_{n+1}+\epsilon} \dot{\phi}(\tau)d\tau=2\pi \tag{5.44}$$

holds. Because of (5.30), (5.33), we obtain

$$\dot{\phi}(\tau)=c+e^{-\gamma(\tau-t_n)}\dot{\chi}(t_n+\epsilon)\,. \tag{5.45}$$

Inserting this relation into (5.44), we arrive at

$$c(t_{n+1}-t_n)+\dot{\chi}(t_n+\epsilon)\frac{1}{\gamma}\left(1-e^{-\gamma(t_{n+1}-t_n)}\right)=2\pi\,, \tag{5.46}$$

which is an equation for (5.41) provided $\dot{\chi}(t_n+\epsilon)$ is known. But $\dot{\chi}$ has been determined in (5.43), so that with (5.41) the relation (5.46) reduces to

$$\Delta=\frac{1}{c}\left(2\pi-\frac{A}{\gamma}\right)\,. \tag{5.47}$$

Clearly, the coupling strength A must be sufficiently small, i.e. $A < 2\pi\gamma$, because otherwise the pulse interval Δ that must be positive would become negative. (We shall study the situation with $A > 2\pi\gamma$ in Sects. 6.12 and 6.13.) If $A > 0$, the mutual coupling between the neurons is excitatory and the pulse intervals decrease, or, in other words, the pulse rate increases, and for $A < 0$, inhibitory coupling, the pulse rate decreases.

Exercise. Solve (5.40) (a) for arbitrary times; and (b) for equidistant times. Hint: (a) put $\dot{\chi}(t_n + \epsilon) = e^{-\gamma t_n} \cdot \eta_n$,
write the resulting equations in the form

$$\eta_{n+1} - \eta_n = a_{n+1}, \quad n = 0, 1, ..., N-1,$$

and sum the l.h.s. and r.h.s., respectively, over n.

(b) $\sum_{n=0}^{N-1} e^{-\alpha n} = \left(1 - e^{-\alpha N}\right)\left(1 - e^{-\alpha}\right)^{-1}$.

5.4 Frequency Pulling and Mutual Activation of Two Neurons

In the preceding section we assumed that the external (sensory) signals, i.e. C_1 and C_2, are equal. What happens if $C_1 \neq C_2$? To study this case, we start from (5.15), (5.17)

$$\ddot{\phi}_1 + \gamma\dot{\phi}_1 = Af(\phi_2) + C_1 \,, \tag{5.52}$$

$$\ddot{\phi}_2 + \gamma\dot{\phi}_2 = Af(\phi_1) + C_2 \,. \tag{5.53}$$

In analogy to (5.30) we make the substitution

$$\phi_j = \frac{1}{\gamma} C_j t + \chi_j = c_j t + \chi_j, \quad j = 1, 2 \tag{5.54}$$

and obtain

$$\ddot{\chi}_1 + \gamma\dot{\chi}_1 = Af(\chi_2 + c_2 t) \,, \tag{5.55}$$

$$\ddot{\chi}_2 + \gamma\dot{\chi}_2 = Af(\chi_1 + c_1 t) \,. \tag{5.56}$$

Because of the cross-wise coupling in (5.55),(5.56), the jump times of $\dot{\chi}_1$ are given by $t_n^{(2)}$ and those of $\dot{\chi}_2$ by $t_n^{(1)}$. Otherwise we may proceed as in Sect. 5.3

and obtain for

$$t_n^{(2)} + \epsilon \leq t \leq t_{n+1}^{(2)} - \epsilon \,, \tag{5.57}$$

$$\dot{\chi}_1(t) = \dot{\chi}_1\left(t_n^{(2)} + \epsilon\right) e^{-\gamma(t - t_n^{(2)})} \,, \tag{5.58}$$

and, correspondingly, for

$$t_n^{(1)} + \epsilon \leq t \leq t_{n+1}^{(1)} - \epsilon \,, \tag{5.59}$$

$$\dot{\chi}_2 = \dot{\chi}_2\left(t_n^{(1)} + \epsilon\right) e^{-\gamma(t - t_n^{(1)})} \,. \tag{5.60}$$

Furthermore we obtain the recursive equations (compare to (5.40))

$$\dot{\chi}_1\left(t_{n+1}^{(2)} + \epsilon\right) = \dot{\chi}_1\left(t_n^{(2)} + \epsilon\right) e^{-\gamma(t_{n+1}^{(2)} - t_n^{(2)})} + A \,, \tag{5.61}$$

$$\dot{\chi}_2\left(t_{n+1}^{(1)} + \epsilon\right) = \dot{\chi}_2\left(t_n^{(1)} + \epsilon\right) e^{-\gamma(t_{n+1}^{(1)} - t_n^{(1)})} + A \,. \tag{5.62}$$

Under the assumption of steady-state conditions, where

$$t_{n+1}^{(1)} - t_n^{(1)} = \Delta_1, \quad t_{n+1}^{(2)} - t_n^{(2)} = \Delta_2 \,, \tag{5.63}$$

and

$$\dot{\chi}_1\left(t_{n+1}^{(2)} + \epsilon\right) = \dot{\chi}_1\left(t_n^{(2)} + \epsilon\right) \equiv \dot{\chi}_1 \,, \tag{5.64}$$

$$\dot{\chi}_2\left(t_{n+1}^{(1)} + \epsilon\right) = \dot{\chi}_2\left(t_n^{(1)} + \epsilon\right) \equiv \dot{\chi}_2 \,, \tag{5.65}$$

we obtain

$$\dot{\chi}_1 \equiv \chi_1\left(t_N^{(2)} + \epsilon\right) = A\left(1 - e^{-\gamma\Delta_2}\right)^{-1} \,, \tag{5.66}$$

$$\dot{\chi}_2 \equiv \chi_2\left(t_N^{(1)} + \epsilon\right) = A\left(1 - e^{-\gamma\Delta_1}\right)^{-1} \,. \tag{5.67}$$

We now have to determine Δ_1 and Δ_2, which, by analogy with (5.44), are defined by

$$\int_{t_n^{(1)}}^{t_{n+1}^{(1)}} \dot{\phi}_1 dt = 2\pi \,, \tag{5.68}$$

$$\int_{t_n^{(2)}}^{t_{n+1}^{(2)}} \dot{\phi}_2 dt = 2\pi \,. \tag{5.69}$$

When evaluating (5.68) and (5.69), we must observe that (5.58) and (5.60), and thus ϕ_1, ϕ_2, are defined only on intervals. To make our analysis as simple as possible (whereby we incidentally capture the most interesting case), we assume

$$| \gamma \Delta_1 | << 1 \, , \quad | \gamma \Delta_2 | << 1 \, . \tag{5.70}$$

Then (5.68) and (5.69) read

$$c_1 \Delta_1 + \dot{\chi}_1 \Delta_1 = 2\pi \, , \tag{5.71}$$

$$c_2 \Delta_2 + \dot{\chi}_2 \Delta_2 = 2\pi \, , \tag{5.72}$$

respectively, which because of (5.66), (5.67) and (5.70) can be transformed into

$$c_1 \Delta_1 + \frac{A}{\gamma} \frac{\Delta_1}{\Delta_2} = 2\pi \, , \tag{5.73}$$

$$c_2 \Delta_2 + \frac{A}{\gamma} \frac{\Delta_2}{\Delta_1} = 2\pi \, . \tag{5.74}$$

Let us discuss these equations in two ways:

1. We may prescribe Δ_1 and Δ_2 and determine those c_1, c_2 (that are essentially the sensory inputs) that give rise to Δ_1, Δ_2.
2. We prescribe c_1 and c_2 and determine Δ_1, Δ_2. Since $\omega_j = 2\pi/\Delta_j$ are the axonal pulse frequencies, we express our results using those

$$\omega_1 = 2\pi \frac{(c_1 2\pi + c_2 A/\gamma)}{4\pi^2 - A^2/\gamma^2} \, , \tag{5.75}$$

$$\omega_2 = 2\pi \frac{(c_1 A/\gamma + c_2 2\pi)}{4\pi^2 - A^2/\gamma^2} \, . \tag{5.76}$$

Their difference and sum are particularly simple

$$\omega_2 - \omega_1 = \frac{c_2 - c_1}{1 + A/(2\pi\gamma)} \, , \tag{5.77}$$

$$\omega_1 + \omega_2 = \frac{c_1 + c_2}{1 - A/(2\pi\gamma)} \, . \tag{5.78}$$

These results exhibit a number of remarkable features of the coupled neurons: According to (5.78) their frequency sum, i.e. their activity, is enhanced by positive coupling A. Simultaneously, according to (5.77) some frequency pulling occurs. According to (5.75), neuron 1 becomes active even for vanishing or negative c_1 (provided $| c_1 2\pi | < c_2 A/\gamma$), if neuron 2 is activated by c_2. This has an important application in the interpretation of the perception of Kaniza figures, and, more generally, to associative memory, as we shall demonstrate below (Sect. 6.4).

5.5 Stability Equations

An important question concerns the stability of the behavior of a system, which means in the present case the stability of the phase-locked state that we derived in the preceding section. To this end, we start with the equations for the dendritic current and the phase of neuron 1 that are subject to the impact of neuron 2. For reasons that will become clear below, we include the fluctuating forces. Thus, we begin with the equations

$$\dot{\psi}_1(t) = -\gamma\psi_1(t) + af(\phi_2(t)) + F_{\psi,1}(t)\,, \tag{5.79}$$

$$\dot{\phi}_1(t) = c\psi_1(t) + p_{\text{ext}}(t) + F_{\phi,1}(t)\,. \tag{5.80}$$

We shall perform the stability analysis in two ways, namely in the conventional mathematical way and in a physically or physiologically realistic manner. Let us start with the first approach, in which case we put

$$F_{\psi,1} = F_{\phi,1} = 0 \quad \text{for all times}\,. \tag{5.81}$$

Then we assume that at time $t = t_0$ a new initial condition

$$\psi_1(t_0) \rightarrow \psi_1(t_0) + \eta_1(t_0) \tag{5.82}$$

is imposed on (5.79) and (5.80) so that in the course of time the dendritic current, and via (5.80) the axonal phase, develop differently than before,

$$\psi_1(t) \rightarrow \psi_1(t) + \eta_1(t)\,, \quad \phi_1(t) \rightarrow \phi_1(t) + \xi_1(t)\,. \tag{5.83}$$

We may also introduce another new initial condition

$$\phi_1(t_0) \rightarrow \phi_1(t_0) + \xi_1(t_0)\,, \tag{5.84}$$

which leads to a new time-development of the phase and dendritic current according to

$$\phi_1(t) \rightarrow \phi_1(t) + \xi_1(t)\,, \quad \psi_1(t) \rightarrow \psi_1(t) + \eta_1(t)\,. \tag{5.85}$$

Also both new initial conditions (5.82) and (5.84) may be imposed simultaneously. We shall speak of an asymptotically stable solution to (5.79) and (5.80) if

$$\text{stable}: \quad \eta_1(t)\,, \xi_1(t) \rightarrow 0 \quad \text{for} \quad t \rightarrow \infty\,, \tag{5.86}$$

and of a marginally stable solution if ξ_1, η_1 remain small if they are initially small. Now let us consider the second approach that we may call the *physiological or physical*. In this case, the perturbations are physically realized and represented by specific fluctuating forces in (5.79) and (5.80). Such a perturbation may be a δ-kick in (5.79)

$$F_{\psi,1} = \eta_1(t_0)\delta(t - t_0)\,, \quad F_{\phi,1} = 0\,. \tag{5.87}$$

The effect of a δ-kick can be dealt with in the by now well-known way, namely by integrating both sides of (5.79) over a small interval around $t = t_0$, which leads to

$$\psi_1(t_0 + \epsilon) = \psi_1(t_0 - \epsilon) + \eta_1(t_0) . \tag{5.88}$$

But (5.88) is nothing but the definition of a new initial condition in accordance with (5.82). What does this result mean to the second equation, i.e. (5.80)? Before the jump this equation reads

$$\dot{\phi}_1(t_0 - \epsilon) = c\psi_1(t_0 - \epsilon) + p_{\text{ext}}(t) , \tag{5.89}$$

whereas after the jump it becomes

$$\dot{\phi}_1(t_0 + \epsilon) = c\,(\psi_1(t_0) + \eta_1(t_0)) + p_{\text{ext}}(t_0) . \tag{5.90}$$

A comparison between (5.90) with (5.89) tells us that a velocity jump of the phase of neuron 1 has happened.

Let us consider the impact of a δ-kick on the phase ϕ_1 according to (5.80). Using

$$F_{\phi,1}(t) = \xi_1(t_0)\delta(t - t_0) , \quad F_{\psi,1} = 0 \tag{5.91}$$

and integrating over the time interval around $t = t_0$ leads to

$$\phi_1(t_0 + \epsilon) = \phi_1(t_0 - \epsilon) + \xi_1(t_0) , \tag{5.92}$$

which means that (5.91) causes a phase jump. As can be seen from (5.80), the same effect can be achieved if the external signal p_{ext} contains a δ-kick. After these preparations, we are in a position to consider an equation from which the dendritic currents have been eliminated. Differentiating (5.80) with respect to time, inserting (5.79) and using (5.80) again, leads to

$$\ddot{\phi}_1 + \gamma\dot{\phi}_1 = Af(\phi_2) + C(t) + F_1(t) . \tag{5.93}$$

Note that this elimination only makes sense if $c \neq 0$ and $a \neq 0$. The quantities occurring on the r.h.s. of (5.93) are defined by

$$C = \gamma p_{\text{ext}} + \dot{p}_{\text{ext}} \tag{5.94}$$

and

$$F_j = \gamma F_{\phi,j} + \dot{F}_{\phi,j} + cF_{\psi,j}, \quad j = 1, 2 . \tag{5.95}$$

The solution of (5.93) requires that the initial conditions are known. From the physiological point of view, we assume that at the initial time $t = 0$ there is no external input, i.e. $p_{\text{ext}}(0) = 0$, and that the neurons are at rest, i.e.

$$\psi_1(0) = 0, \quad \phi_1(0) = 0 . \tag{5.96}$$

Furthermore a look at (5.80) tells us that at that time also

$$\dot{\phi}_1(0) = p_{\text{ext}}(0) + f(\phi_2(0)) = 0 \tag{5.97}$$

holds. Equations (5.96) and (5.97) serve as initial conditions for the integration of (5.93). We assume p_{ext} is in the form

$$p_{\text{ext}}(t) = p\left(1 - e^{-\gamma t}\right), \tag{5.98}$$

which guarantees that the external signal vanishes at time $t = 0$ and reaches its steady-state value after times bigger than $1/\gamma$. We integrate both sides of (5.93) over time from $t = 0$ to t and obtain

$$\dot{\phi}_1(t) + \gamma\phi_1(t) = A\int_0^t f(\phi_2(\sigma))d\sigma + \gamma pt + B_1(t) \tag{5.99}$$

with the abbreviation

$$B_j(t) = \int_0^t F_j(\sigma)d\sigma, \quad j = 1, 2. \tag{5.100}$$

A different time constant Γ instead of γ in (5.98) will lead to an additive constant in (5.99), which can be compensated by a time shift or correspondingly by a shift of phase. Since (5.98) is used both in (5.99) and the corresponding one in which the indices 1 and 2 are interchanged, this time is the same for both neurons and does not change their relative phase but only the position of the absolute phase. Thus, phase locking is preserved. Since we want to study the behavior of deviations from the phase-locked state, we must also consider its equation

$$\dot{\phi}(t) + \gamma\phi(t) = A\int_0^t f(\phi(\sigma))d\sigma + \gamma pt. \tag{5.101}$$

In the following we want to show how different kinds of perturbations give rise to different kinds of stability behaviors. We study the combined impact of (5.87) and (5.91). We now wish to derive the basic equations for our stability analysis. To this end, we compare the perturbed time-evolution of $\phi_1(t)$ according to (5.99) with the unperturbed evolution of ϕ according to (5.101). We insert the hypothesis

$$\phi_j(t) = \phi + \xi_j(t), \quad j = 1, 2, \tag{5.102}$$

into (5.99) and substract (5.101), which yields

$$\dot{\xi}_1 + \gamma\xi_1 = A\left(\int_0^t f(\phi(\sigma) + \xi_2(\sigma))d\sigma - \int_0^t f(\phi(\sigma))d\sigma\right) + B_1(t). \tag{5.103}$$

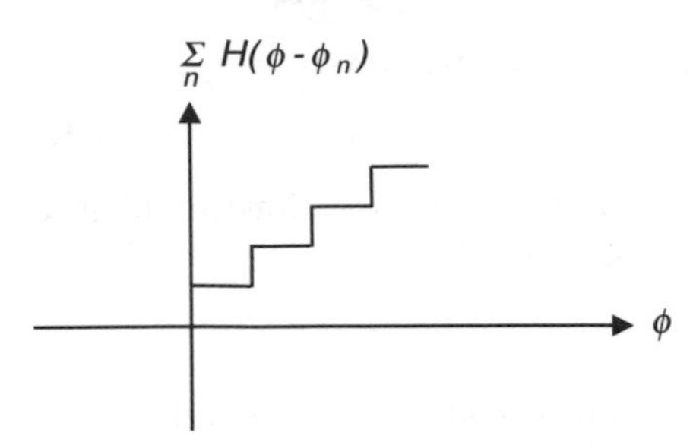

Fig. 5.11. The sum (5.105) of Heaviside functions

The following analysis is a little bit intricate, so we refer the reader who is not interested in too many mathematical details to the final result which is presented below in (5.116) and (5.117). On the other hand, the more mathematically interested reader surely will have read Chap. 4 and in that Sects. 4.1 and 4.2, so we can capitalize on his or her knowledge. The integrals occurring on the r.h.s. of (5.103) are well-known to us. Using the explicit form of $f(\phi)$, we obtain

$$\int_0^t f(\phi)dt \equiv \int_0^t \dot{\phi} \sum_n \delta(\phi - \phi_n)dt = \sum_n \int^{\phi} \delta(\phi - \phi_n)d\phi = J(\phi) \,, \quad (5.104)$$

which is nothing but a sum over step functions or Heaviside functions (4.15), which is depicted in Fig. 5.11,

$$J(\phi) = \sum_n H(\phi - \phi_n) \,, \quad (5.105)$$

where

$$\begin{aligned} H(\sigma) &= 0 \quad \text{for} \quad \sigma < 0 \,, \\ &= 1/2 \quad \text{for} \quad \sigma = 0 \,, \\ &= 1 \quad \text{for} \quad \sigma > 0 \,. \end{aligned} \quad (5.106)$$

We now can immediately use the results that we have derived in Sect. 4.2 starting from $G(T)$ defined in (4.40). In order to apply our former results to the present case, we have to use the identifications

$$\phi(t) \to \phi(t) - 2\pi n \,, \quad \xi(t) \to \xi_2(T) \,, \quad (5.107)$$

$$t \to \sigma, \quad T \to t \,, \quad (5.108)$$

and to sum (4.40) over the integers n. Thus, we have to make the substitution

$$G(t) \to G_{\text{tot}}(t) = \sum_n \Bigg(\int_0^t \delta(\phi(\sigma) + \xi_2(\sigma) - 2\pi n) \left(\dot{\phi}(\sigma) + \dot{\xi}_2(\sigma)\right) d\sigma - \int_0^t \delta(\phi(\sigma) - 2\pi n)\dot{\phi}(\sigma) d\sigma \Bigg) \,. \quad (5.109)$$

Furthermore in accordance with the definitions of the times defined the Sect. 4.2, we use the identifications

$$t_0^- \to t_n^- : \quad \phi\left(t_n^-\right) + \xi_2\left(t_n^-\right) = 2\pi n \tag{5.110}$$

$$t_0 \to t_n : \quad \phi(t_n) = 2\pi n \,. \tag{5.111}$$

Since the time difference between t_n and t_n^- is small, we may use the replacement

$$\xi_2\left(t_n^-\right) \approx \xi_2(t_n) \,. \tag{5.112}$$

In this way, the relation (4.51) can be translated into

$$\left(t_n - t_n^-\right) = \xi_2(t_n)/\dot{\phi}(t_n) \,, \tag{5.113}$$

where for the steady phase-locked state we may use

$$\dot{\phi}(t_n) = \dot{\phi}(t_0) \,. \tag{5.114}$$

Finally we note that after (4.55) we recognized that G practically has the properties of a δ-function provided the time interval (5.113) is small enough and we take care of the corresponding area. Thus, our final result reads

$$G_{\text{tot}}(t) \approx \sum_n \delta(t - 2\pi n)\xi_2(t_n)/\dot{\phi}(t_0) \,. \tag{5.115}$$

It should be noted that the result (5.115) can be achieved in several ways, but the present is probably the most concise approach. The reader who was not so much interested in mathematical trickery can now resume reading this section, namely inserting (5.115) into (5.103), and we arrive at our final result

$$\dot{\xi}_1 + \gamma\xi_1 = aD(t)\xi_2 + B_1(t) \,, \tag{5.116}$$

and by an exchange of indices 1,2

$$\dot{\xi}_2 + \gamma\xi_2 = aD(t)\xi_1 + B_2(t) \,, \tag{5.117}$$

where

$$a = A\dot{\phi}^{-1} \quad \text{and} \quad D(t) = \sum_\ell \delta(t - t_\ell) \,, \tag{5.118}$$

where t_ℓ is defined by $\phi(t_\ell) = 2\pi\ell$. Adding or subtracting the equations (5.116) and (5.117) to or from each other, we obtain

$$\dot{Z} + \gamma Z = aD(t)Z + B_+ \,, \tag{5.119}$$

where

$$Z = \xi_1 + \xi_2, \quad B_+ = B_1 + B_2 \tag{5.120}$$

and

$$\dot{\xi} + \gamma\xi = -aD(t)\xi + B(t) \,, \tag{5.121}$$

where

$$\xi = \xi_2 - \xi_1, \quad B = B_2 - B_1 \,. \tag{5.122}$$

5.6 Phase Relaxation and the Impact of Noise

In the present section we want to bring (5.121) to life by studying the effects of various perturbations B on the phase difference $\xi = \xi_2 - \xi_1$. To this end, we evaluate B_1, B_2 (5.100) with F_j (5.95) for the combined impact of the δ-perturbations (5.87) and (5.91). Because of the properties of the δ-functions, we obtain

$$B_j(t) = \gamma\xi_j(t_0)H(t-t_0) + \xi_j(t_0)\delta(t-t_0) + c\eta_1(t_0)H(t-t_0)\,. \qquad (5.123)$$

This expression is the sum of two kinds of functions with different time-dependencies, namely $\delta(t)$ and $H(t)$. Let us discuss the corresponding terms separately

$$1)\quad B = \xi_0\delta(t-t_0)\,. \qquad (5.124)$$

Such a B is caused by a fluctuation

$$F_{\phi,j}(t) = \xi_j(t_0)\delta(t-t_0)\,, \qquad (5.125)$$

and a *correlated perturbation* $F_{\psi,j}$ so that

$$\gamma\xi_j(t_0) + c\eta_1(t_0) = 0\,. \qquad (5.126)$$

If not otherwise stated, later in this book we shall assume that (5.126) is fulfilled. The effect of (5.124) can be dealt with by solving (5.121) as an initial value problem, whereby B may be dropped. As we shall show, $\xi(t) \to 0$ for $t \to \infty$, i.e. the phase-locked state is stable, at least for $\mid a \mid$ small enough.

$$2)\quad B = \begin{cases} 0 & \text{for } t < t_0 \\ B_0 = \text{const.} & \text{for } t \geq t_0\,. \end{cases} \qquad (5.127)$$

This behavior occurs if in (5.123) the fluctuation $F_{\psi,j} \propto \delta(t-t_0)$ is used. As a result, to be demonstrated below, the relative phase is changed by a constant amount. This indicates neutral (marginal) stability.

3) B represents noise, where

$$< B(t) > = 0,\ < B(t)B(t') > = Q\delta(t-t')\,. \qquad (5.128)$$

As a result, the relative phase shows finite fluctuations (i.e. *no* phase diffusion).

Let us treat these cases in detail.

1) $B = \xi_0\delta(t-t_0)$.
In this case, we may treat (5.121) with $B \equiv 0$ subject to the initial value $\xi(t_0) = \xi_0$ (and $\xi(t) = 0$ for $t < t_0$)

$$\dot{\xi}(t) + \gamma\xi(t) = -a\sum\delta(t-t_n)\xi(t)\,, \qquad (5.129)$$

where a is defined by (5.118). Because the phase ϕ refers to the steady state, a is a constant. We first study the solution of (5.129) in the interval

$$t_n < t < t_{n+1} \tag{5.130}$$

and obtain

$$\xi(t) = \xi(t_n + \epsilon)e^{-\gamma(t-t_n)} . \tag{5.131}$$

At times t_n we integrate (5.129) over a small interval around t_n and obtain

$$\xi(t_n + \epsilon) = \xi(t_n - \epsilon) - a\xi(t_n - \epsilon) . \tag{5.132}$$

Since ξ undergoes a jump at time t_n, there is an ambiguity with respect to the evaluation of the last term in (5.132). Instead of $t_n - \epsilon$ we might equally well choose $t_n + \epsilon$ or an average over both expressions. Since we assume, however, that a is a small quantity, the error is of higher order and we shall, therefore, choose ξ at $t_n - \epsilon$ as shown in (5.132). [Taking the average amounts to replacing $(1 - a)$ with $(1 - a/2)/(1 + a/2)$.] On the r.h.s. of (5.132), we insert (5.131) for $t = t_n + \epsilon$ and thus obtain

$$\xi(t_n + \epsilon) = (1 - a)\xi(t_{n-1} + \epsilon)e^{-\gamma(t_n-t_{n-1})} . \tag{5.133}$$

Since the t_n's are equally spaced, we put

$$t_n - t_{n-1} \equiv \Delta . \tag{5.134}$$

For the interval

$$t_N < t < t_{N+1} \tag{5.135}$$

the solution reads

$$\xi(t) = \xi(t_0 + \epsilon)(1 - a)^N e^{-\gamma\Delta\cdot N-\gamma(t-t_N)} . \tag{5.136}$$

Since for excitatory interaction, $a > 0$, and $\dot{\phi}$ sufficiently large the absolute value of $1 - a$ is smaller than unity, (5.136) shows that the phase deviation $\xi(t)$ relaxes towards zero in the course of time. We turn to the cases 2) and 3), where $B(t)$ is time-dependent and the equation to be solved reads

$$\dot{\xi}(t) + \gamma\xi(t) = B(t) - a\sum \delta(t - t_n)\xi(t) . \tag{5.137}$$

In the interval

$$t_{n-1} < t < t_n \tag{5.138}$$

the general solution of (5.137) reads

$$\xi(t) = \xi(t_{n-1} + \epsilon)e^{-\gamma(t-t_{n-1})} + \int_{t_{n-1}}^{t} e^{-\gamma(t-\sigma)}B(\sigma)d\sigma . \tag{5.139}$$

We first treat the case that $B(t)$ is non-singular. At time t_n, the integration of (5.137) over a small time interval yields

$$\xi(t_n + \epsilon) = \xi(t_n - \epsilon) - a\xi(t_n - \epsilon)\,. \tag{5.140}$$

We put $t = t_n - \epsilon$ in (5.139) and thus obtain

$$\xi(t_n - \epsilon) = \xi(t_{n-1} + \epsilon)e^{-\gamma(t_n - t_{n-1})} + \int\limits_{t_{n-1}}^{t_n} e^{-\gamma(t_n - \sigma)} B(\sigma) d\sigma\,. \tag{5.141}$$

We replace the r.h.s. of (5.140) by means of (5.141) and obtain

$$\xi(t_n + \epsilon) = (1 - a)\left\{\xi(t_{n-1} + \epsilon)e^{-\gamma\Delta} + \hat{B}(t_n)\right\}, \tag{5.142}$$

where we abbreviated the integral in (5.141) by $\hat{B}$. Introducing the variable x instead of ξ, we can rewrite (5.142) in an obvious manner by means of

$$x_n = (1 - a)\left\{x_{n-1}e^{-\gamma\Delta} + \hat{B}_n\right\}. \tag{5.143}$$

To solve the set of equations (5.143), we make the substitution

$$x_n = \left((1 - a)e^{-\gamma\Delta}\right)^n y_n \tag{5.144}$$

and obtain a recursion formula for y_n,

$$y_n - y_{n-1} = (1 - a)^{-n+1} e^{\gamma t_n} \hat{B}_n\,. \tag{5.145}$$

Summing over both sides of (5.145), we obtain

$$\sum_{n=1}^{N} (y_n - y_{n-1}) = \sum_{n=1}^{N} (1 - a)^{-n+1} e^{\gamma t_n} \hat{B}_n\,, \tag{5.146}$$

or, written more explicitly,

$$y_N = y_0 + \sum_{n=1}^{N} (1 - a)^{-n+1} \int\limits_{t_{n-1}}^{t_n} e^{\gamma\sigma} B(\sigma) d\sigma\,. \tag{5.147}$$

By means of (5.144), we obtain the final result in the form (with $t_n - t_{n-1} = \Delta$)

$$\begin{aligned} x_N &= y_0 \left((1 - a)e^{-\gamma\Delta}\right)^N \\ &\quad + \sum_{n=1}^{N} (1 - a)^{N-n+1} e^{-\gamma\Delta N} \int\limits_{t_{n-1}}^{t_n} e^{\gamma\sigma} B(\sigma) d\sigma\,. \end{aligned} \tag{5.148}$$

So far our treatment was quite general. Now we have to treat the cases 2) and 3) separately.

2) $B = B_0$ for all times. B is time-independent, the integral in (5.148) can immediately be evaluated

$$x_N = y_0\left((1-a)e^{-\gamma\Delta}\right)^N + \sum_{n=1}^{N}(1-a)^{N-n+1}e^{-\gamma\Delta N}B\frac{1}{\gamma}\left(e^{\gamma t_n} - e^{\gamma t_{n-1}}\right) . \tag{5.149}$$

The sum is a geometric series, and we obtain

$$x_N = x_0(1-a)^N e^{-\gamma\Delta N} + \frac{B}{\gamma}\left(e^{\gamma\Delta} - 1\right) \cdot \frac{1-(1-a)^N e^{-\gamma\Delta N}}{(1-a)e^{\gamma\Delta} - 1} . \tag{5.150}$$

For $N \to \infty, x_N$ acquires a constant value, i.e. the phase shift persists and becomes for $a \ll 1$ particularly simple, namely $x_N = B/\gamma$.

3) We now turn to the case in which B is a stochastic function with the properties shown in (5.128). In the following we shall study the correlation function for the case N large, and

$$\mid N - N' \mid \text{ finite} . \tag{5.151}$$

Using (5.148), the correlation function can be written in the form

$$< x_N x_{N'} > = \sum_{n=1}^{N}\sum_{n'=1}^{N'}(1-a)^{N-n+1}e^{-\gamma\Delta N}(1-a)^{N'-n'+1} \times e^{-\gamma\Delta N'}\int_{t_{n-1}}^{t_n}\int_{t_{n'-1}}^{t_{n'}} d\sigma d\sigma' \cdot e^{\gamma\sigma}e^{\gamma\sigma'} < B(\sigma)B(\sigma') > . \tag{5.152}$$

We evaluate (5.152) in the case

$$N' \geq N \tag{5.153}$$

and use the fact that B is δ-correlated with strength Q. Then (5.152) acquires the form

$$R \equiv < x_N x_{N'} > = \sum_{n=1}^{N}(1-a)^{N-n+1}e^{-\gamma\Delta N}(1-a)^{N'-n+1} e^{-\gamma\Delta N'}\frac{1}{2\gamma}\left(e^{2\gamma t_n} - e^{2\gamma t_{n-1}}\right)Q . \tag{5.154}$$

The evaluation of the sum is straightforward and yields

$$R = \frac{Q}{2\gamma}\left(e^{2\gamma\Delta} - 1\right)\left\{(1-a)^{-2}e^{2\gamma\Delta} - 1\right\}^{-1}(1-a)^{N'-N}e^{-\gamma\Delta(N'-N)} , \tag{5.155}$$

which for

$$a << 1 , \tag{5.156}$$

can be written as

$$R = e^{-(\gamma\Delta + a)(N'-N)} \cdot \frac{Q}{2\gamma} . \tag{5.157}$$

The correlation function has the same form as we would expect from a purely continuous treatment of (5.137), i.e. in which the δ-functions are smeared out.

5.7 Delay Between Two Neurons

We treat the case in which the interaction between the neurons has a time delay τ. We shall assume that both the deviation ξ from the phase-locked state and ϕ are delayed by the same amount τ. The crucial interaction term replacing that of (5.103) then reads

$$A \int_0^t f(\phi(t'-\tau) + \xi_2(t'-\tau))dt' - A \int_0^t f(\phi(t'-\tau))dt' . \tag{5.158}$$

Its effect can be reduced to that without delay by using the formal replacements

$$\xi_j(\sigma) \to \xi_j(\sigma - \tau) , \quad j = 1, 2, \quad \phi(\sigma) \to \phi(\sigma - \tau) . \tag{5.159}$$

In this way we are immediately led to the equation

$$\dot{\xi}(t) + \gamma\xi(t) = -a \sum_n \delta(t - t_n)\xi(t - \tau) \tag{5.160}$$

by replacing $\xi(t)$ with $\xi(t-\tau)$ on the r.h.s. of (5.121). In this section, we ignore fluctuations, i.e. we put $B = 0$. Again we assume that ϕ undergoes a stationary process. We write

$$t_n = n\Delta \tag{5.161}$$

and assume that τ is an integer multiple of Δ (in Chap. 6 we shall relax this assumption),

$$\tau = M\Delta , \quad M \quad \text{integer} . \tag{5.162}$$

In the following we capitalize on the fact that we may convert the (delay) *differential* equation (5.160) into a (delay) *difference* equation (see (5.165) below). In a way that is by now well-known from the treatment of (5.121) in Sect. 5.6, we obtain the jump condition

$$\xi(t_n + \epsilon) = \xi(t_n - \epsilon) - a\xi(t_{n-M} - \epsilon) \tag{5.163}$$

and the solution in between jumps

$$\xi(t_n - \epsilon) = e^{-\gamma\Delta}\xi(t_{n-1} + \epsilon) . \tag{5.164}$$

A combination of (5.163) and (5.164) yields

$$\xi(t_n + \epsilon) = e^{-\gamma\Delta}\xi(t_{n-1} + \epsilon) - ae^{-\gamma\Delta}\xi(t_{n-M-1} + \epsilon) . \tag{5.165}$$

We use the notation $x_n = \xi(t_n + \epsilon)$ so that (5.165) reads

$$x_n = e^{-\gamma\Delta}x_{n-1} - ae^{-\gamma\Delta}x_{n-M-1} . \tag{5.166}$$

In order to solve this recursion equation, we make the hypothesis

$$x_n = \beta^n x_0 . \tag{5.167}$$

Inserting (5.167) into (5.166) yields

$$\beta^{M+1} - e^{-\gamma\Delta}\beta^M + ae^{-\gamma\Delta} = 0 . \tag{5.168}$$

Under the assumption that a is a small quantity, the roots of (5.168) can be written in the form

$$\beta_1 = e^{-\gamma\Delta}\left(1 - ae^{M\gamma\Delta}\right) . \tag{5.169}$$

and

$$\beta_{j+1} = a^{1/M}e^{2\pi ij/M}, \quad j = 1, .., M . \tag{5.170}$$

The general solution to (5.166) is a superposition of the solutions (5.167) with the corresponding eigenvalues β, i.e.

$$x_n = \beta_1^n x_{01} + \beta_2^n x_{02} + ... + \beta_{M+1}^n x_{0M+1} , \tag{5.171}$$

where the coefficients x_{0j} are still unknowns. In order to determine them, we must invoke the initial conditions. Without delay, a single value, $x_0 \equiv \xi_0$ will suffice to start the iteration described by the recursion relation (5.166). With $M > 0$ the situation is more involved, however. To illustrate it, let us consider the recursion equation (5.166) for $M = 1$ and let us start with $n = 1, 2, 3$.

$$x_1 = e^{-\gamma\Delta}(x_0 - ax_{-1}) , \tag{5.172}$$

$$x_2 = e^{-\gamma\Delta}(x_1 - ax_0) , \tag{5.173}$$

$$x_3 = e^{-\gamma\Delta}(x_2 - ax_1) . \tag{5.174}$$

Obviously, to start the iteration, we must know both x_0 and x_{-1}. They are fixed by the initial conditions $x_0 = \xi_0, x_{-1} = \xi_{-1}$, where ξ_0, ξ_{-1} are prescribed. On the other hand, we know the general form of the solution

$$x_n = \beta_1^n x_{01} + \beta_2^n x_{02} . \tag{5.175}$$

To determine the unknown x_{01}, x_{02}, we require

$$n = 0: \; x_0 = x_{01} + x_{02} = \xi_0 \,, \tag{5.176}$$

$$n = -1: \; x_{-1} = \beta_1^{-1} x_{01} + \beta_2^{-1} x_{02} = \xi_{-1} \,, \tag{5.177}$$

which are two linear equations for these. We are now in a position to present the general case, $M \geq 1$. Equations (5.172)–(5.174) are generalized to

$$\begin{aligned}
n = 1: &\qquad x_1 = e^{-\gamma \Delta} \left(x_0 - a x_{-M} \right) \\
n = 2: &\qquad x_2 = e^{-\gamma \Delta} \left(x_1 - a x_{-M+1} \right) \\
&\vdots \\
n = M: &\qquad x_M = e^{-\gamma \Delta} \left(x_{M-1} - a x_{-1} \right) \\
n = M+1: &\qquad x_{M+1} = e^{-\gamma \Delta} \left(x_M - a x_0 \right) .
\end{aligned} \tag{5.178}$$

In this general case, to start the iteration, $x_0, x_{-1}, ..., x_{-M}$ must be given by the initial values $\xi_0, \xi_{-1}, ..., \xi_{-M}$, respectively. Using (5.171), this requirement leads to $M+1$ linear equations for the $M+1$ unknowns $x_{0j}, j = 1, ..., M+1$.

The solution is straightforward, but gives rise to rather lengthy formulae, at least in the general case. Rather, what concerns us is the relaxation behavior of the deviation $\xi(t_n) = x_n$. β_1 is smaller than unity, and equally all the other roots (5.170) have an absolute value that is smaller than unity, though the roots (5.170) indicate oscillatory behavior. All in all we may state that the absolute value of all roots is smaller than unity so that phase-locking occurs even in the case of delay.

5.8 An Alternative Interpretation of the Lighthouse Model

In Sect. 5.5 we derived a basic equation that allowed us to study the impact of noise as well as phase relaxation. For later purposes, we mention that (5.99) with (5.100) can be interpreted in quite a different manner. Let us again study the linearized case; then one may readily convince oneself that (5.99) can be considered as a result of the elimination of the dendritic currents ψ_1 from the following equations

$$\text{neuron 1 axon 1}: \qquad \dot{\phi}_1 + \gamma \phi_1 = c \psi_1 + F_{\phi 1} \,, \tag{5.179}$$

$$\text{dendrite 1}: \qquad \dot{\psi}_1 = a f(\phi_2) + (\gamma p + \dot{p})/c + F_{\psi 1} \,. \tag{5.180}$$

Indeed, (5.180) can be solved in the form

$$\psi_1 = \int_0^t a f(\phi(\sigma)) d\sigma + \int_0^t (\gamma p + \dot{p})/c d\sigma + \int_0^t F_{\psi 1} d\sigma \,, \tag{5.181}$$

which, whcn inserted into (5.179) yields our former (5.99). When the fluctuating force $F_{\phi j}$ is chosen as a δ-function and $F_{\psi j} = 0$ we have to deal with the initial value problem for the phases ϕ_j. It is not difficult to find a nonlinear version of (5.179), namely

$$\dot{\phi}_1 + \gamma \phi_1 = cS(\psi_1) + F_{\phi 1} \,. \tag{5.182}$$

For neuron 2, the equations can be obtained by exchanging the indices 1 and 2. Clearly, the neural dynamics described by (5.179) and (5.180) differs from that of Sect. 5.1. From the mathematical point, however, the impact of initial conditions and their equivalence to a δ-function perturbation can be more easily seen and utilized.

6. The Lighthouse Model. Many Coupled Neurons

This chapter represents, besides Chap. 8, the main part of my book. Throughout, Chap. 6 deals with an arbitrary number N of neurons, i.e. large neural nets are treated. In Sect. 6.1 I formulate the basic equations that include arbitrary couplings between neurons, consisting of dendrites, synapses and axons. Both arbitrary delays and fluctuations are taken care of. The following sections focus on the existence and stability of the phase-locked state. I adopt a pedagogic style by stepwise increasing the complications. Sections 2–4 neglect delays and fluctuations. They show under which conditions a phase-locked state is possible. This requires, in particular, the same sensory inputs for the neurons. I study also the impact of different inputs. This leads to the concept of associative memory (Sect. 6.4). The following Sects. 6.5–6.10 include the effect of delays on associative memory, and, in particular, on the phase-locked state, its stability and its stability limits. The example of two delay times will be presented in detail, but general results are also included. Spatial aspects, namely phase-waves, are dealt with in Sect. 6.9. The remaining part of this chapter deals with two still more involved topics. Section 6.11 studies the combined action of fluctuations and delay. Sections 6.12 and 6.13 are devoted to the strong coupling limits where the formerly treated phase-locked state becomes unstable.

6.1 The Basic Equations

Since the network equations are a straightforward generalization of those of two neurons, formulated in Sect. 5.1, we can write them down immediately. The generation of a dendritic current $\psi_m(t)$ of dendrite m is described by

$$\dot{\psi}_m(t) = \sum_k a_{mk} P_k(t - \tau_{mk}) - \gamma \psi_m(t) + F_{\psi,m}(t) . \tag{6.1}$$

The coefficients a_{mk} are called synaptic strengths, $P_k(t-\tau)$ are pulses of axon k with delay time τ_{mk}, γ is the dendritic damping, and $F_{\psi,m}(t)$ is a stochastic force. A more general relation between axonal pulses and dendritic currents is given by

$$\psi_m(t) = \sum_k \int_{-\infty}^{t} G_{mk}(t,\sigma) P_k(\sigma - \tau_{mk}) d\sigma + \tilde{F}_m(t) \,. \tag{6.2}$$

The kernel G_{mk} may contain fluctuations that result, for example, from the failure of the opening of vesicles. In this book, we will not pursue this version (6.2), however. Again, by means of a periodic function f with δ-peaks, P_k is expressed by a phase ϕ_k (see (6.9) below). The phase ϕ_k is subject to the equation

$$\dot{\phi}_k(t) = S\left(\sum_m c_{km}\psi_m(t-\tau'_{km}) + p_{\mathrm{ext,k}}(t-\tau''_{km}), \Theta_k\right) + F_{\phi,k}(t) \,. \tag{6.3}$$

$S(X)$ is a sigmoid function that may be chosen in the Naka–Rushton form, $p_{\mathrm{ext,k}}$ is an external signal, Θ_k is a threshold, and $F_{\phi,k}$ is a fluctuating force. The coefficients c_{km} are assumed time-independent, and τ' and τ'' are delay times.

In the following we assume that the neural net operates in the (practically) linear regime of S, so S in (6.3) can be replaced by its argument. (A proportionality factor can be absorbed in $c_{km}, p_{\mathrm{ext,k}}$, respectively.) It is a simple matter to eliminate ψ_m from (6.1) and (6.3). To this end we differentiate (6.3) with respect to time, and replace $\dot{\psi}_m$ on the r.h.s. of the resulting equation by the r.h.s. of (6.1). Finally we use again (6.3) to eliminate ψ_m. The reader is advised to perform these steps using paper and pencil. Changing the notation of indices, we thus finally obtain a set of equations of a rather simple structure

$$\ddot{\phi}_j(t) + \gamma\dot{\phi}_j(t) = \sum_{\ell,m} A_{j\ell;m} P_\ell(t-\tau_{j\ell m}) + C_j(t) + \hat{F}_j(t) \,. \tag{6.4}$$

The reader must not be shocked, however, by the rather lengthy expressions of the abbreviations. We will hardly need their explicit form.

$$A_{j\ell;m} = c_{jm} a_{m\ell} \,, \tag{6.5}$$

$$\tau_{j\ell m} = \tau'_{jm} + \tau_{m\ell} \,, \tag{6.6}$$

$$C_j(t) = \gamma \sum_m p_{\mathrm{ext,j}}\left(t - \tau''_{jm}\right) + \sum_m \dot{p}_{\mathrm{ext,j}}\left(t - \tau''_{km}\right) , \tag{6.7}$$

$$\hat{F}_j(t) = \gamma F_{\phi,j}(t) + \sum_m c_{jm} F_{\psi,m}\left(t - \tau'_{jm}\right) + \dot{F}_{\phi,j}(t) \,. \tag{6.8}$$

We will use the explicit form, already known to us,

$$P_\ell(t) = f\left(\phi_\ell(t)\right) = \dot{\phi}_\ell(t) \sum_n \delta(\phi_\ell(t) - 2\pi n) \,. \tag{6.9}$$

The initial conditions in (6.4) are

$$\dot{\phi}_j(0) = \phi_j(0) = 0 . \tag{6.10}$$

6.2 A Special Case. Equal Sensory Inputs. No Delay

To get a first insight into the action of many coupled neurons, we first treat a special case of (6.4). We assume that all sensory inputs are equal

$$C_j = C \quad \text{for all} \quad j , \tag{6.11}$$

all thresholds equal

$$\Theta_j = \Theta \quad \text{for all} \quad j , \tag{6.12}$$

and p_{ext} and Θ_j time-independent, i.e.

$$C(t) : \text{ constant} . \tag{6.13}$$

No delay implies (see 6.6)

$$\tau_{j\ell m} = 0 . \tag{6.14}$$

Equation (6.4) then reads

$$\ddot{\phi}_j(t) + \gamma\dot{\phi}_j(t) = \sum_\ell A_{j\ell} P_\ell(t) + C + \hat{F}_j(t) . \tag{6.15}$$

Note that in the absence of time delays the index m is superfluous and has been dropped.

Let us study the possibility of phase-locking (or more precisely, synchrony). In this case

$$\phi_j(t) = \phi(t) \quad \text{for all} \quad j . \tag{6.16}$$

In a first step, we ignore noise, i.e. we put $\hat{F}_j = 0$. Inserting (6.16) into (6.15) and using (6.9) yields

$$\ddot{\phi}(t) + \gamma\dot{\phi}(t) = \sum_\ell A_{j\ell} P(\phi(t)) + C . \tag{6.17}$$

This equation provides us with a necessary condition for phase locking: Since the l.h.s. of (6.17) is independent of the index j, so must be the r.h.s.! This implies

$$\sum_\ell A_{j\ell} = A \quad \text{independent of} \quad j . \tag{6.18}$$

This condition might look rather restrictive, but below we will convince ourselves that it can be fulfilled in realistic networks.

When using (6.18) in (6.17), we discover that the resulting equation for the phase-locked state is well-known to us and was solved in Sect. 5.3. Thus we are left with the stability analysis of the phase-locked state, including the effect of noise. By analogy with Sect. 5.5 we integrate (6.15) over time and observe the initial conditions (6.10), and use the explicit form of P_k (6.9). Because of

$$\int_0^t \sum_n \delta(\phi_k(t) - 2\pi n)\dot{\phi}_k(t)dt = \int_0^{\phi_k(t)} \sum_n \delta(\phi - 2\pi n)d\phi \equiv J(\phi_k(t)) \quad (6.19)$$

we obtain

$$\dot{\phi}_j(t) + \gamma\phi_j(t) = \sum_k A_{jk}J(\phi_k(t)) + Ct + B_j(t)\,, \quad (6.20)$$

where

$$B_j(t) = \int_0^t \hat{F}_j(\sigma)d\sigma\,. \quad (6.21)$$

The function (6.19) is plotted in Fig. 5.11. The equation of the phase-locked state reads

$$\dot{\phi}(t) + \gamma\phi(t) = AJ(\phi(t)) + Ct\,, \quad (6.22)$$

where (6.18) is assumed.

To study the impact of noise and the stability of the phase-locked state, we put

$$\phi_j = \phi + \xi_j \quad (6.23)$$

and subtract (6.22) from (6.20), which yields

$$\dot{\xi}_j + \gamma\xi_j = \sum_k A_{jk}\left\{\int_0^{\phi+\xi_k} \sum_n \delta(\phi - 2\pi n)d\phi - \int_0^{\phi} \sum_n \delta(\phi - 2\pi n)d\phi\right\} + B_j(t)\,, \quad (6.24)$$

where we use the explicit form of J.

In order to study stability and/or if the noise sources F are small, we may assume that ξ_j is a small quantity. According to basic rules of analysis, we may evaluate the curly bracket by replacing the first integral by its argument, multiplied by ξ_k. When proceeding from a δ-function containing ϕ in its argument to one that contains time t, a correction factor $\dot{\phi}^{-1}$ must be added. This had been derived in Sect. 4.1 so that we can use those results here. Thus, eventually, we obtain

$$\dot{\xi}_j(t) + \gamma\xi_j(t) = D(t)\sum_k a_{jk}\xi_k(t) + B_j(t)\,, \quad (6.25)$$

where

$$D(t) = \sum_{\ell=-\infty}^{+\infty} \delta\left(t - t_\ell^+\right) , \tag{6.26}$$

where t_ℓ^+ is defined by $\phi(t_\ell^+) = 2\pi\ell, \ell$ an integer. The coefficients a_{jk} are defined by $a_{jk} = A_{jk}\dot{\phi}^{-1}$, where $\dot{\phi}(t_\ell^+)$ is independent of ℓ because of the stationarity of ϕ. Readers who have carefully read Sect. 5.5, surely noted that here I just presented an alternative derivation of (5.116) and (5.117). This may help to get more familiar with the δ-formalism.

The set of linear differential equations (6.25) can be solved by the standard procedure. We introduce eigenvectors $\mathbf{v}^\mu$ with components v_j^μ and eigenvalues λ_μ so that

$$\sum_j v_j^\mu a_{jk} = \lambda_\mu v_k^\mu \tag{6.27}$$

and put

$$\sum_j v_j^\mu \xi_j(t) = \eta_\mu(t) , \tag{6.28}$$

$$\sum_j v_j^\mu B_j(t) = \tilde{B}_\mu(t) . \tag{6.29}$$

This allows us to transform (6.25) into the uncoupled equations for the collective variables η_μ

$$\dot{\eta}_\mu + \gamma\eta_\mu = D(t)\lambda_\mu\eta_\mu + \tilde{B}_\mu(t) . \tag{6.30}$$

We thus obtain the remarkable result that the whole net reacts to fluctuations as if it is composed of independent, self-coupled neurons subject to collective fluctuating forces $\tilde{B}_\mu(t)$. Thus we may directly transfer the results of Sect. 5.6 to the present case. The stability of the phase-locked state is guaranteed (sufficient condition) if the real parts of all eigenvalues λ_μ are negative.

6.3 A Further Special Case. Different Sensory Inputs, but No Delay and No Fluctuations

The starting point of our considerations is again the set of (6.4) with $\tau_{j\ell m} = 0$. Thus we wish to deal with the equations

$$\ddot{\phi}_j(t) + \gamma\dot{\phi}_j(t) = \sum_\ell A_{j\ell}P_\ell(t) + C_j . \tag{6.31}$$

Note that in contrast to the preceding section we do not (yet) impose conditions on the $A_{j\ell}$s. We put

$$\dot{\phi}_j(t) = x_j(t) + c_j \equiv x_j(t) + C_j/\gamma \tag{6.32}$$

and use (6.9) to obtain equations for x_j

$$\dot{x}_j + \gamma x_j = \sum_\ell A_{j\ell} f(\phi_\ell) . \tag{6.33}$$

Since (6.33) is, at least from a formal point of view, a linear equation in x_j, we make the hypothesis

$$x_j = \sum_\ell x_j^{(\ell)} \tag{6.34}$$

and require

$$\dot{x}_j^{(\ell)} + \gamma x_j^{(\ell)} = A_{j\ell} f(\phi_\ell) . \tag{6.35}$$

For the solution of these equations, we proceed by close analogy to Sects. 5.3 and 5.4. We first assume that the times at which the spikes of the δ-functions in $f(\phi_\ell)$ occur are given quantitites. We denote them by $t_{N(\ell)}$, because to each $f(\phi_\ell)$ there belongs a specific ℓ-dependent time series. By analogy to Sects. 5.3 and 5.4 we assume equidistant jumps, i.e.

$$t_{N(\ell)+1} - t_{N(\ell)} = \Delta_\ell \tag{6.36}$$

and steady states. Then the solution of (6.35) reads (see (5.66))

$$x_j^{(\ell)}(t_{N(\ell)} + \epsilon) = A_{j\ell} \left(1 - e^{-\gamma \Delta_\ell}\right)^{-1} , \tag{6.37}$$

or for an arbitrary time with $t_{N(\ell)} + \epsilon < T < t_{N(\ell)+1} - \epsilon$

$$x_j^{(\ell)}(T) = e^{-\gamma(T - t_{N(\ell)})} A_{j\ell} \left(1 - e^{-\gamma \Delta_\ell}\right)^{-1} . \tag{6.38}$$

Using (6.34), we obtain the final result

$$x_j(T) = \sum_\ell e^{-\gamma(T - t_{N(\ell)})} A_{j\ell} \left(1 - e^{-\gamma \Delta_\ell}\right)^{-1} . \tag{6.39}$$

The jump intervals are determined by

$$\int_{t_{N(\ell)}}^{t_{N(\ell)+1}} \dot{\phi}_\ell(\sigma) d\sigma = 2\pi . \tag{6.40}$$

In order to evaluate the integral in (6.40), we use (6.32) and (6.39), where under the assumption

$$\gamma \left(t_{N(\ell)+1} - t_{N(\ell)}\right) << 1 \tag{6.41}$$

(6.39) can be approximated by

$$x_j = \sum_{\ell'} A_{j\ell'}/(\gamma\Delta_{\ell'}) \,. \tag{6.42}$$

Thus we obtain (generalizing (5.73) and (5.74))

$$c_\ell \Delta_\ell + \Delta_\ell \sum_{\ell'} A_{\ell\ell'}/(\gamma\Delta_{\ell'}) = 2\pi \,. \tag{6.43}$$

These equations relate the axonal pulse frequencies $\omega_\ell = 2\pi/\Delta_\ell$ to the strengths of the sensory inputs, c_ℓ. The corresponding equations for ω_ℓ are linear and read

$$c_\ell + \sum_{\ell'} [A_{\ell\ell'}/(2\pi\gamma)]\omega_{\ell'} = \omega_\ell \,. \tag{6.44}$$

They can be solved under the usual conditions for linear equations. Here we want to discuss two aspects of particular relevance for neural nets. In general, we may expect that all pulse frequencies ω_ℓ are different. Since equal frequencies are a prerequisite for phase locking, this phenomenon will be absent. On the other hand, we may derive from (6.44) a sufficient condition for equal frequencies, $\omega_\ell = \omega$, namely

$$\omega = c_\ell \left(1 - \sum_{\ell'} A_{\ell\ell'}/(2\pi\gamma)\right)^{-1} \,. \tag{6.45}$$

This equation implies that its r.h.s. is independent of the index ℓ. This is for instance the case if

$$c_\ell = c, \quad \text{and} \quad \sum_{\ell'} A_{\ell\ell'} \quad \text{independent of} \quad \ell \,. \tag{6.46}$$

These were the conditions for the phase-locked state in Sect. 6.2.

A second important aspect is this: Depending on the coupling coefficients $A_{\ell\ell'}$, even those ω_ℓ may become non-zero, for which $c_\ell = 0$. This may give rise to (weak) associative memory as we will demonstrate in the next section. Note that in all cases discussed here only those solutions are allowed for which $\omega_\ell \geq 0$ for all ℓ. This imposes limitations on c_ℓ and $A_{\ell\ell'}$.

6.4 Associative Memory and Pattern Filter

Let us first discuss what is meant by "associative memory". Such a memory serves the completion of data. A telephone book is an example. When we look up the name "Alex Miller", this book provides us with his telephone number. Or when we see the face of a person whom we know, our brain tells us (or should tell us!) his or her name. The results of the preceding section may

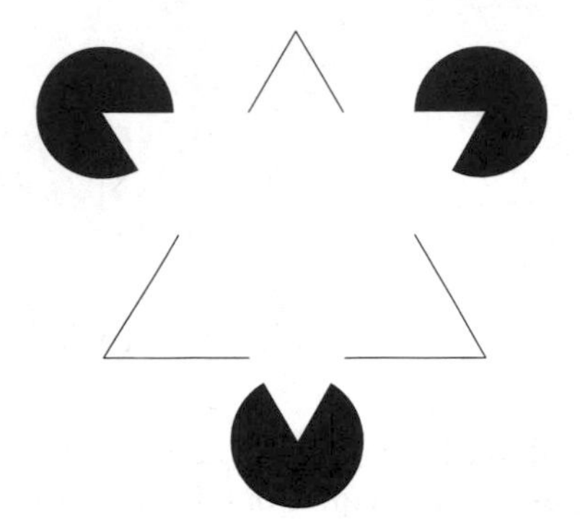

Fig. 6.1. A Kanisza figure. A white triangle seems to float above a black triangle

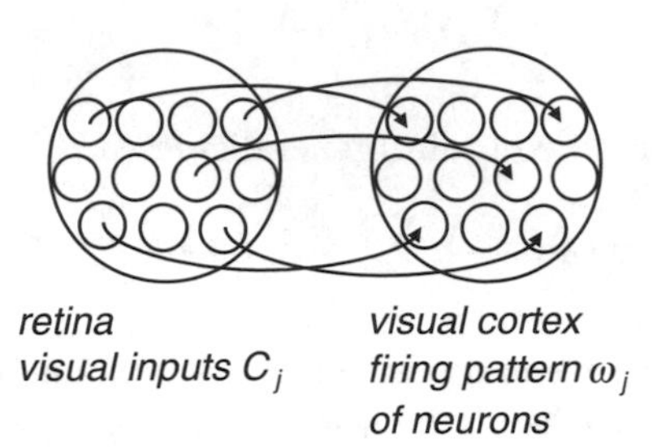

Fig. 6.2. Mapping from sensory cells of the retina onto neurons of the visual cortex

serve as a model that allows us to understand a particular kind of associative memory for which an example is provided by looking at Fig. 6.1. In spite of the interruptions of the lines, we see a complete triangle! By means of the results of the preceding section, this phenomenon can be understood as follows, whereby we use a highly schematic representation: When we look at a picture, our retina receives input signals that are eventually transmitted to the visual cortex, where they act as sensory inputs C_j on neuron j. Because of these inputs and their mutual coupling, the neurons "fire" with specific axonal pulse rates, so that the input pattern C (cf. Fig. 6.2) is translated into a "firing" pattern ω.

Let us assume that the neural net has learned preferred patterns, for instance an *uninterrupted* line. To make things simple, consider a retina consisting of only three sensory cells and the neural net also consisting of only three neurons. An uninterrupted line will then be represented by the activation scheme (1,1,1), whereas an interrupted scheme is represented by (1,0,1). Can a network be constructed, i.e. can the A_{jk}s be chosen in such a way that the sensory input vector $\mathbf{c} = \mathbf{C}/\gamma \equiv (c_1, c_2, c_3) = (1,0,1)$ is practically converted into a vector $\boldsymbol{\omega} \equiv (\omega_1, \omega_2, \omega_3)$ whose components

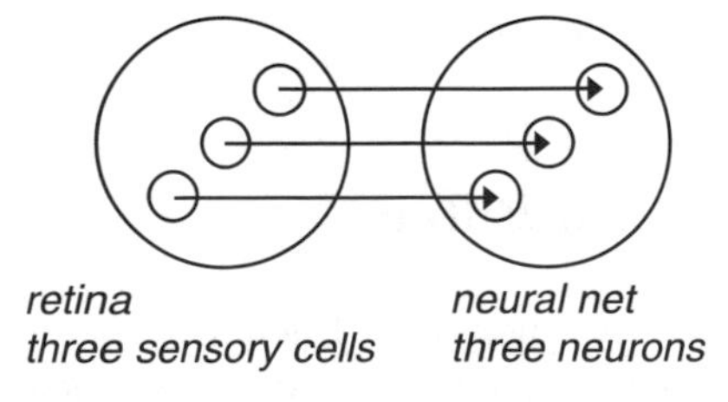

Fig. 6.3. Same as Fig. 6.2, but only three sensory cells and three neurons are involved

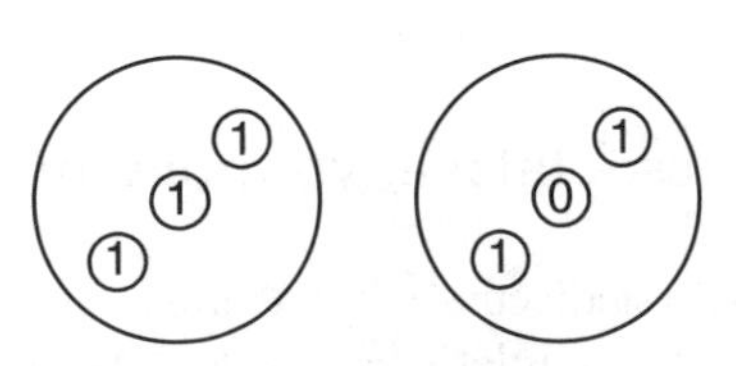

Fig. 6.4. *L.h.s.*: stored prototype pattern, an uninterrupted line; *r.h.s.*: pattern received by retina

are (approximately) equal, i.e. represent an uninterrupted line? We choose $\tilde{A}_{\ell\ell'} = A_{\ell\ell'}/(2\pi\gamma)$ according to the following scheme:

$$\begin{pmatrix} \tilde{A}_{11} & \tilde{A}_{12} & \tilde{A}_{13} \\ \tilde{A}_{21} & \tilde{A}_{22} & \tilde{A}_{23} \\ \tilde{A}_{31} & \tilde{A}_{32} & \tilde{A}_{33} \end{pmatrix} = \begin{pmatrix} 0,9 & 0 & 0 \\ 0,8 & 0,1 & 0 \\ 0,8 & 0 & 0,1 \end{pmatrix} . \tag{6.47}$$

The solution of the corresponding equations (6.43) then reads

$$\begin{aligned} &\boldsymbol{\omega} = (\omega_1, \omega_2, \omega_3) \approx (10, 9, 10)\,, \quad \text{i.e.} \\ &\boldsymbol{\omega} \approx 10 \cdot (1, 1, 1)\,. \end{aligned} \tag{6.48}$$

In other words, the neural nets "sees" the middle point in spite of the fact that its corresponding original cell on the retina is *not* excited. The way we wrote down the matrix scheme so as to obtain the required result (6.48) may seem rather mysterious. Therefore we present the general theory lying behind that construction. The reader may easily check the general formulas using our above explicit example. Thereby we use a few results from algebra. We denote those patterns that are preferably recognized by the neural net as prototype vectors $\mathbf{v}_\mu$ that are distinguished by an index μ. To be precise, we write each $\mathbf{v}_\mu$ as a column vector

$$\mathbf{v}_\mu = \begin{pmatrix} v_{\mu 1} \\ v_{\mu 2} \\ \vdots \\ v_{\mu N} \end{pmatrix} . \tag{6.49}$$

Similarly, the sensory input vector $\mathbf{c}$ is written as

$$\mathbf{c} = \begin{pmatrix} c_1 \\ c_2 \\ \vdots \\ c_N \end{pmatrix} . \tag{6.50}$$

In our above example,

$$\mathbf{c} = \begin{pmatrix} 1 \\ 0 \\ 1 \end{pmatrix} \tag{6.51}$$

is such a vector, whereas

$$\boldsymbol{\omega} = \mathbf{v} = \begin{pmatrix} 1 \\ 1 \\ 1 \end{pmatrix} \tag{6.52}$$

is a prototype vector of the axonal pulse rates. We first assume that the "synaptic strengths" $\tilde{A}_{\ell\ell'}$ are given, and that the prototype vectors $\mathbf{v}_\mu$ are just the eigenvectors to the matrix $\left(\tilde{A}_{\ell\ell}\right)$ with eigenvalues λ_μ. (We assume for simplicity that all λ_μ values are different from each other.) Thus in vector notation

$$\tilde{A}\mathbf{v}_\mu = \lambda_\mu \mathbf{v}_\mu \,. \tag{6.53}$$

Now consider (6.44) that reads in vector notation

$$(1 - \tilde{A})\boldsymbol{\omega} = \mathbf{c} \,. \tag{6.54}$$

For our procedure we need one more step, namely the adjoint vectors $\mathbf{v}_\nu^+, \nu = 1, ..., N$. They are defined as row vectors

$$\mathbf{v}_\nu^+ = \left(v_{\nu 1}^+, v_{\nu 2}^+, ..., v_{\nu N}^+ \right) , \tag{6.55}$$

and must obey the relations

$$\left(\mathbf{v}_\nu^+ \mathbf{v}_\mu \right) = \delta_{\nu\mu} \,, \tag{6.56}$$

i.e. the scalar product between the two vectors must equal the Kronecker δ,

$$\delta_{\nu\mu} = \begin{cases} 1 \text{ for } \nu = \mu \\ 0 \text{ for } \nu \neq \mu \end{cases} . \tag{6.57}$$

By means of $\mathbf{v}_\mu, \mathbf{v}_\nu^+$ we can easily construct the relationship between $\boldsymbol{\omega}$ and $\mathbf{c}$. We decompose $\boldsymbol{\omega}$ and $\mathbf{c}$ into the $\mathbf{v}_\mu$:

$$\boldsymbol{\omega} = \sum_\mu \alpha_\mu \mathbf{v}_\mu \,, \tag{6.58}$$

$$\mathbf{c} = \sum_\mu \beta_\mu \mathbf{v}_\mu \,, \tag{6.59}$$

where the coefficients α_μ, β_μ are given by means of (see the exercises)

$$\alpha_\mu = \left(\mathbf{v}_\mu^+ \boldsymbol{\omega} \right) , \tag{6.60}$$

$$\beta_\mu = \left(\mathbf{v}_\mu^+ \mathbf{c} \right) . \tag{6.61}$$

Inserting (6.58) and (6.59) into (6.54), using (6.53) and multiplying both sides by $\mathbf{v}_\nu^+$ (see the exercises) yields

$$(1 - \lambda_\nu)\alpha_\nu = \beta_\nu \,, \tag{6.62}$$

$$\alpha_\nu = (1 - \lambda_\nu)^{-1} \beta_\nu \,. \tag{6.63}$$

Inserting this result into (6.58) yields

$$\boldsymbol{\omega} = \sum_\nu (1 - \lambda_\nu)^{-1} \mathbf{v}_\nu \,. \tag{6.64}$$

This means that, depending on the size of the eigenvalues λ_ν, the $\mathbf{v}_\nu$s are mapped onto the axonal pulse frequencies with quite different weights $(1-\lambda_\nu)^{-1}$. In our above simple example we made the choice

$$\mathbf{v}_1 = \begin{pmatrix} 1 \\ 1 \\ 1 \end{pmatrix}, \quad \mathbf{v}_2 = \begin{pmatrix} 0 \\ 1 \\ 0 \end{pmatrix}, \quad \mathbf{v}_3 = \begin{pmatrix} 0 \\ 0 \\ 1 \end{pmatrix} \tag{6.65}$$

with $\lambda_1 = 0,9, \quad \lambda_2 = 0,1, \quad \lambda_3 = 0,1.$

But how were we able to choose the coefficients $\tilde{A}_{\ell\ell'}$? In fact, there is an algebraic formula, namely

$$\tilde{A}_{\ell\ell'} = \sum_\mu \lambda_\mu v_{\mu\ell} v^+_{\mu\ell'} \,. \tag{6.66}$$

The $\mathbf{v}^+_\mu$ that obey (6.56) can easily be calculated and are

$$\mathbf{v}_1^+ = (1,0,0), \quad \mathbf{v}_2^+ = (-1,1,0), \quad \mathbf{v}_3^+ = (-1,0,1)\,. \tag{6.67}$$

As we may note, we can construct pulse-coupled neural networks so that their axonal pulse patterns come very close to specific, learned preferred patterns – though there remains a residuum of other small patterns superimposed on the preferred pattern. We call this effect *weak associative memory.* A number of hypotheses on how the human brain learns patterns have been established. These hypotheses imply, for instance, that our brain learns those visual patterns that are frequently seen. There is, however, another important aspect that relates to the concept Gestalt. Namely in interpreting forms, we prefer to perceive "good Gestalts" (good shapes). But why are, for example, straight lines, or triangles or circles "good Gestalts"? This is, in this author's opinion, an unsolved problem.

A last comment: Is the completion of a line just in our imagination, or do neurons fire that correspond to the gap in the line. Experiments on monkeys show that the latter is the case.

In this section our theoretical treatment of weak associative memory has been based on a simplified model. In Sect. 6.5 we want to show that the underlying model can be much more general.

Exercises. (1) Derive (6.60) and (6.61).
Hint: Start from (6.58) and (6.59) and use (6.56).
(2) Derive (6.62), (6.63).
Hint: Use (6.57), (6.60) and (6.61).

6.5 Weak Associative Memory. General Case*

In this section we will see that the basic equations (6.44) that relate the pulse frequencies $\boldsymbol{\omega}$ with the sensory inputs can be derived from the general equations (6.4)

$$\ddot{\phi}_j(t) + \gamma\dot{\phi}_j(t) = \sum_{\ell,m} A_{j\ell;m} P_\ell\left(t - \tau_{j\ell m}\right) + C_j(t) + \hat{F}_j(t)\,, \tag{6.68}$$

where all the quantities have been defined in Sect. 6.1. The initial conditions in (6.68) are

$$\dot{\phi}_j(0) = \phi_j(0) = 0\,. \tag{6.69}$$

In spite of the highly nonlinear character of (6.68) because of P_ℓ, and the delay times, we can derive an exact relationship between the pulse rates of the axons j, where we use time-averaged pulse rates. We will assume that $C_j(t)$ is practically constant over an averaging interval of duration T and that T is large enough to cover the delay times τ. We integrate (6.68) over the time interval T and divide both sides by T

$$\begin{aligned} &T^{-1}\left(\dot{\phi}_j(T+t_0) - \dot{\phi}_j(t_0) + \gamma\phi_j(T+t_0) - \gamma\phi_j(t_0)\right) \\ &= T^{-1}\sum_{\ell,m} A_{j\ell;m} \int_{t_0}^{t_0+T} P_\ell(t-\tau_{j\ell m})dt \\ &+ T^{-1}\int_{t_0}^{t_0+T} C_j(t)dt + T^{-1}\int_{t_0}^{t_0+T} \hat{F}_j(t)dt\,. \end{aligned} \tag{6.70}$$

We discuss the individual terms in the mathematical limit $T \to \infty$. (For all practical purposes, T can be chosen finite but large enough.) We obtain

$$T^{-1}(\dot{\phi}(T+t_0) - \dot{\phi}(t_0)) \to 0\,, \tag{6.71}$$

because the phase velocity is finite. Since $\frac{1}{2\pi}\left(\phi_j(T+t_0) - \phi_j(t_0)\right)$ is the number of rotations of the phase angle ϕ_j,

$$\frac{1}{2\pi}T^{-1}(\phi_j(T+t_0) - \phi_j(T)) = \omega_j \tag{6.72}$$

is the pulse rate. Since

$$\int_{t_0}^{t_0+T} P_\ell(t-\tau)dt \tag{6.73}$$

contains δ-peaks according to (6.9), we obtain

$$(6.73) = \int_{t_0}^{t_0+T} \sum_n \delta\left(\phi_\ell(t-\tau) - 2\pi n\right)\dot{\phi}_\ell(t-\tau)dt \tag{6.74}$$

$$= \int\limits_{t_0-\tau}^{t_0+T-\tau} \sum_n \delta\left(\phi_\ell(t) - 2\pi n\right) \dot{\phi}_\ell(t) dt \tag{6.75}$$

$$= \int\limits_{\phi_\ell(t_0-\tau)}^{\phi_\ell(t_0+T-\tau)} \sum_n \delta(\phi_\ell - 2\pi n) d\phi_\ell \,, \tag{6.76}$$

which is equal to the number of pulses of axon ℓ in the interval $[\phi_\ell(t_0 - \tau), \phi_\ell(t_0 + T - \tau)]$.

Thus when we divide (6.73) by T and assume steady-state conditions, we obtain the pulse rate

$$T^{-1} \int\limits_{t_0}^{t_0+T} P_\ell(t-\tau) d\tau = \omega_\ell \,. \tag{6.77}$$

For slowly varying or constant $C_j(t) \approx C_j$, we obtain

$$T^{-1} \int\limits_{t_0}^{t_0+T} C_j(t) dt = C_j(t_0) \,. \tag{6.78}$$

Because we assume that the fluctuating forces possess an ensemble average

$$< \hat{F}_j(t) > = 0 \,, \tag{6.79}$$

and under the further assumption that the time average equals the ensemble average we find

$$T^{-1} \int\limits_{t_0}^{t_0+T} \hat{F}_j(t) dt = 0 \,. \tag{6.80}$$

Lumping the results (6.71), (6.72), (6.77), (6.78) and (6.80) together and dividing all terms by $(2\pi\gamma)$, we obtain the *linear* equations

$$\omega_j = \sum_{\ell,m} A_{j\ell;m} (2\pi\gamma)^{-1} \omega_\ell + c_j(t) \,, \tag{6.81}$$

where

$$c_j(t) = (2\pi\gamma)^{-1} C_j(t) \,. \tag{6.82}$$

The equations (6.81) can be solved under the usual conditions. Depending on the coupling coefficients $A_{j\ell} = \sum_m A_{j\ell;m}$, even those ω_ℓ may become nonzero for which $c_\ell = 0$. On the other hand, only those solutions are allowed for which $\omega_\ell \geq 0$ for all ℓ. This imposes limitations on c_ℓ and $A_{\ell\ell'}$. Putting $\sum_m A_{j\ell;m} = A_{j\ell}$, we recover (6.44) which formed the basis of our discussion in Sect. 6.4.

6.6 The Phase-Locked State of N Neurons. Two Delay Times

In this section and the following we resume our study of the solutions of the basic equation (6.4) (see also (6.68)), but without time averages. Our main goal will be the study of the impact of delays on the spiking behavior of the net. Since this section is more mathematical than the previous sections (except Sect. 6.5), we summarize the essential results. In this section we show how the phase-locked state can be calculated. More importantly, the subsequent section shows that delays diminish the stability of the phase-locked state. Furthermore, after a perturbation the phases of the individual axonal pulses relax towards the phase-locked state in an oscillatory fashion.

In order to demonstrate our general approach, we consider N neurons, two delay times τ_1, τ_2 and no fluctuations. The phase equations (6.4) acquire the form

$$\ddot{\phi}_j(t) + \gamma\dot{\phi}_j(t) = \sum_{k,\ell} A_{jk,\ell} f(\phi_k(t-\tau_\ell)) + C_j \,, \tag{6.83}$$

where we assume

$$C_j = C = \text{const.} \tag{6.84}$$

We assume the existence of the phase-locked state in which case $\phi_j = \phi$ for $j = 1, ..., N$. ϕ obeys

$$\ddot{\phi}(t) + \gamma\dot{\phi}(t) = A_1 f(\phi(t-\tau_1)) + A_2 f(\phi(t-\tau_2)) + C \,, \tag{6.85}$$

where

$$A_1 = \sum_k A_{jk,1}, A_2 = \sum_k A_{jk,2} \tag{6.86}$$

are assumed to be independent of j. This condition is by no means unrealistic as we will demonstrate below. We treat the case where ϕ is periodic in time with a period Δ still to be determined. The spikes of $f(\phi(t))$ are assumed to occur at times $t_n = n\Delta$, n an integer. We thus have

$$t - \tau_j = t_{n_j} = n_j\Delta, \quad j = 1, 2 \,, \tag{6.87}$$

or, after solving (6.87) with respect to t,

$$t = n_j\Delta + \tau_j = n\Delta + \tau'_j = t_n + \tau'_j \,, \tag{6.88}$$

where τ' is assumed to obey the inequality

$$0 \le \tau'_j < \Delta \,. \tag{6.89}$$

We introduce the new variable x by means of

$$\dot{\phi} = c + x(t); \quad c = C/\gamma \tag{6.90}$$

so that (6.85) is transformed into

$$\dot{x}(t) + \gamma x(t) = A_1 f(\phi(t - \tau_1)) + A_2 f(\phi(t - \tau_2)). \tag{6.91}$$

In the following we first assume that the r.h.s. is a given function of time t. Because of the δ-function character of f, we distinguish between the following four cases, where we incidentally write down the corresponding solutions of (6.91)

$$\text{I}: \quad t_n + \tau_1' < t < t_n + \tau_2' : x(t) = e^{-\gamma(t - t_n - \tau_1)} x(t_n + \tau_1' + \epsilon), \tag{6.92}$$

$$\text{II}: \quad t_n + \tau_2' \mp \epsilon : x(t_n + \tau_2' + \epsilon) = x(t_n + \tau_2' - \epsilon) + A_2, \tag{6.93}$$

$$\text{III}: \quad t_n + \tau_2' < t < t_{n+1} + \tau_1' : x(t) = e^{-\gamma(t - t_n - \tau_2')} x(t_n + \tau_2' + \epsilon), \tag{6.94}$$

$$\text{IV}: \quad t_{n+1} + \tau_1' \pm \epsilon : x(t_{n+1} + \tau_1' + \epsilon) = x(t_{n+1} + \tau_1' - \epsilon) + A_1. \tag{6.95}$$

Combining the results (6.92)–(6.95), we find the following recursion relation

$$x(t_{n+1} + \tau_1' + \epsilon) = e^{-\gamma \Delta} x(t_n + \tau_1' + \epsilon) + e^{-\gamma(\Delta + \tau_1' - \tau_2')} A_2 + A_1. \tag{6.96}$$

Under the assumption of a steady state, we may immediately solve (6.96) and obtain

$$x(t_{n+1} + \tau_1' + \epsilon) = \left(1 - e^{-\gamma \Delta}\right)^{-1} \left(A_1 + e^{-\gamma(\Delta + \tau_1' - \tau_2')} A_2\right). \tag{6.97}$$

The only unknown quantity is Δ. To this end, we require, as usual,

$$\int_{t_n}^{t_{n+1}} \dot{\phi} dt = 2\pi, \tag{6.98}$$

i.e. that ϕ increases by 2π. In order to evaluate (6.98) by means of (6.91), we start from (6.91), which we integrate on both sides over time t

$$\int_{t_n}^{t_{n+1}} (\dot{x}(t) + \gamma x(t)) dt = \int_{t_n}^{t_{n+1}} \left(A_1 f(\phi(t - \tau_1)) + A_2 f(\phi(t - \tau_2))\right) dt \tag{6.99}$$

Because of the steady-state assumption, we have

$$\int_{t_n}^{t_{n+1}} \dot{x}(t) dt = x(t_{n+1}) - x(t_n) = 0 \tag{6.100}$$

so that (6.99) reduces to

$$\gamma \int_{t_n}^{t_{n+1}} x(t)dt = A_1 + A_2 . \tag{6.101}$$

Using this result as well as (6.90) in (6.98), we obtain

$$c\Delta + (A_1 + A_2)/\gamma = 2\pi \tag{6.102}$$

which can be solved for the time interval Δ to yield

$$\Delta = \frac{1}{c}\left(2\pi - (A_1 + A_2)/\gamma\right) . \tag{6.103}$$

We can also determine the values of $x(t)$ in the whole interval by using the relations (6.92)–(6.95). Because the time interval Δ must be positive, we may suspect that $\Delta = 0$ or, according to (6.103),

$$(2\pi - (A_1 + A_2)/\gamma) = 0 \tag{6.104}$$

represents the stability limit of the stationary phase-locked state. We will study this relationship in Sect. 6.10.

6.7 Stability of the Phase-Locked State. Two Delay Times*

In order to study this problem, we assume as initial conditions

$$\dot{\phi}_j(0) = \phi_j(0) = 0 \tag{6.105}$$

and integrate (6.4) over time, thus obtaining

$$\dot{\phi}_j(t) + \gamma\phi_j(t) = \sum_{k,\ell} A_{jk,\ell} J(\phi_k(t - \tau_\ell)) + C_j t + B_j(t) , \tag{6.106}$$

where J has been defined in (6.19). We include the fluctuating forces, put

$$B_j(t) = \int_0^t \hat{F}_j(\sigma)d\sigma , \tag{6.107}$$

and assume that C_j in (6.7) is time-independent. The phase-locked state obeys

$$\dot{\phi}(t) + \gamma\phi(t) = \sum_{k\ell} A_{jk,\ell} J(\phi(t - \tau_\ell)) + Ct . \tag{6.108}$$

In order to study the stability of the state $\phi(t)$, we make the hypothesis

$$\phi_j = \phi + \xi_j , \quad j = 1, ..., N . \tag{6.109}$$

Subtracting (6.108) from (6.106) and assuming

$$C_j = C , \tag{6.110}$$

we obtain

$$\begin{aligned} \dot{\xi}_j(t) + \gamma \xi_j(t) = \sum_{k\ell} & A_{jk,\ell}(J(\phi(t-\tau_\ell) + \xi_j(t-\tau_\ell)) \\ & -J(\phi(t-\tau_\ell))) + B_j(t) . \end{aligned} \tag{6.111}$$

A somewhat extended analysis analogous to that of Sect. 6.2 shows that for small ξ_k these equations reduce to

$$\dot{\xi}_j(t) + \gamma \xi_j(t) = \sum_{k\ell} a_{jk;\ell} D_\ell(t) \xi_k(t) + B_j(t) . \tag{6.112}$$

The quantities on the r.h.s. of (6.112) are defined as follows:

$$D_\ell(t) = \sum_n \delta\left(t - \tau_\ell - t_n^+\right) , \tag{6.113}$$

$$\phi\left(t_n^+\right) = 2\pi n , \tag{6.114}$$

$$a_{jk;\ell} = A_{jk;\ell}\left[\dot{\phi}\left(t_n^+ + \tau_\ell\right)\right]^{-1} . \tag{6.115}$$

To allow for a concise treatment of (6.112), we introduce the matrix $(a_{jk;\ell})$

$$\tilde{A}_\ell = (a_{jk;\ell}) \quad \begin{matrix} j = 1, ..., N \\ k = 1, ..., N \end{matrix} \tag{6.116}$$

so that we cast (6.112) into the vector equation

$$\dot{\boldsymbol{\xi}}(t) + \gamma \boldsymbol{\xi}(t) = \sum_\ell \tilde{A}_\ell D_\ell(t) \boldsymbol{\xi}(t) + \mathbf{B}(t) . \tag{6.117}$$

Again, as before, we introduce the following times for $\ell = 1, 2$

$$\begin{gathered} \tau_\ell = M_\ell \Delta + \tau_\ell' ; \quad 0 \le \tau_\ell' < \Delta ; \\ (n_\ell + M_\ell)\Delta + \tau_\ell' = n\Delta + \tau_\ell' ; \quad n_\ell + M_\ell = n ; \quad n_\ell = n - M_\ell . \end{gathered} \tag{6.118}$$

Under this definition we have the following regions for which we solve (6.117), namely

$\mathrm{I} : n\Delta + \tau_1' < t < n\Delta + \tau_2'$:

$$\boldsymbol{\xi}(t) = e^{-\gamma(t-(n\Delta+\tau_1'))} \boldsymbol{\xi}(n\Delta + \tau_1' + \epsilon) + \hat{\mathbf{B}}(t, n\Delta + \tau_1') , \tag{6.119}$$

where

$$\hat{\mathbf{B}}(t,t') = \int_{t'}^{t} e^{-\gamma(t-\sigma)}\mathbf{B}(\sigma)d\sigma\,, \tag{6.120}$$

$\mathrm{II}: n\Delta + \tau_2' \pm \epsilon:$

$$\boldsymbol{\xi}(n\Delta + \tau_2' + \epsilon) - \boldsymbol{\xi}(n\Delta + \tau_2' - \epsilon) = \tilde{A}_2\boldsymbol{\xi}((n - M_2)\Delta)\,, \tag{6.121}$$

$\mathrm{III}: n\Delta + \tau_2' < t < (n+1)\Delta + \tau_1':$

$$\boldsymbol{\xi}(t) = e^{-\gamma(t-(n\Delta+\tau_2'))}\boldsymbol{\xi}(n\Delta + \tau_2') + \hat{\mathbf{B}}(t, n\Delta + \tau_2')\,, \tag{6.122}$$

$\mathrm{IV}: (n+1)\Delta + \tau_1' \pm \epsilon:$

$$\boldsymbol{\xi}((n+1)\Delta + \tau_1' + \epsilon) - \boldsymbol{\xi}((n+1)\Delta + \tau_1' - \epsilon) = \tilde{A}_1\boldsymbol{\xi}((n+1-M_1)\Delta). \tag{6.123}$$

By means of (6.119)–(6.123) and eliminating the intermediate steps, we obtain the fundamental recursion equation

$$\begin{aligned}\boldsymbol{\xi}((n+1)\Delta + \tau_1' + \epsilon) &= e^{-\gamma\Delta}\boldsymbol{\xi}(n\Delta + \tau_1' + \epsilon) \\ &+ e^{-\gamma(\Delta+\tau_1'-\tau_2')}\tilde{A}_2\boldsymbol{\xi}((n - M_2)\Delta) \\ &+ \tilde{A}_1\boldsymbol{\xi}((n+1-M_1)\Delta) + \hat{\mathbf{B}}((n+1)\Delta + \tau_1'; n\Delta + \tau_1').\end{aligned} \tag{6.124}$$

In order to express the solutions in between the time steps, we use

$$\boldsymbol{\xi}(n\Delta + \tau_1' + \epsilon) = e^{-\gamma\tau_1'}\boldsymbol{\xi}(n\Delta + \epsilon)\,. \tag{6.125}$$

Inserting (6.125) into (6.124), we readily obtain

$$\begin{aligned}\boldsymbol{\xi}_{n+1} = e^{-\gamma\Delta}\boldsymbol{\xi}_n + e^{-\gamma(\Delta-\tau_2')}\tilde{A}_2\boldsymbol{\xi}_{n-M_2} + e^{\gamma\tau_1}\tilde{A}_1\boldsymbol{\xi}_{n+1-M_1} \\ + e^{\gamma\tau_1'}\hat{B}((n+1)\Delta + \tau_1'; n\Delta + \tau_1')\,,\end{aligned} \tag{6.126}$$

where we used the abbreviation

$$\boldsymbol{\xi}(n\Delta + \epsilon) = \boldsymbol{\xi}_n\,. \tag{6.127}$$

Equation (6.126) is valid, if $M_2 \geq 0$ and $M_1 \geq 1$. If $M_1 = 0$, a further recursive step must be performed. To a good approximation we may replace

$$\boldsymbol{\xi}_{n+1-M_1} \quad \text{by} \quad e^{-\Delta\gamma}\boldsymbol{\xi}_{n-M_1}\,. \tag{6.128}$$

In order to solve the difference equations (6.126), we must observe the initial conditions. As is well-known from delay differential equations, these conditions must be defined on a whole interval. In the present case we may capitalize on the fact that we are dealing with difference equations so that we need to consider only a discrete set of initial conditions. On the other hand, we need a *complete set*. In order to demonstrate how to proceed, we first treat the special case of only *one* delay time $\tau = M\Delta$, M an integer, and ignore the fluctuating forces. The equation to be discussed is of the form

$$\boldsymbol{\xi}_{n+1} = e^{-\gamma\Delta}\boldsymbol{\xi}_n + A\boldsymbol{\xi}_{n-M} , \tag{6.129}$$

where A is a matrix. Since we want to consider the process starting at $n = 0$, we need the $M + 1$ initial values

$$\boldsymbol{\xi}_0, \boldsymbol{\xi}_{-1}, ..., \boldsymbol{\xi}_{-M} . \tag{6.130}$$

This then allows us to initiate the solution of the $M + 1$ recursive equations

$$n = 0: \quad \boldsymbol{\xi}_1 = e^{-\gamma\Delta}\boldsymbol{\xi}_0 + A\boldsymbol{\xi}_{-M} , \tag{6.131}$$

$$n = 1: \quad \boldsymbol{\xi}_2 = e^{-\gamma\Delta}\boldsymbol{\xi}_1 + A\boldsymbol{\xi}_{1-M} , \tag{6.132}$$

$$\vdots$$

$$n = M: \quad \boldsymbol{\xi}_{M+1} = e^{-\gamma\Delta}\boldsymbol{\xi}_M + A\boldsymbol{\xi}_0 . \tag{6.133}$$

In the general case of (6.126), we can proceed correspondingly where the initial conditions are given by (6.130) where

$$M = \max(M_1 - 1, M_2) . \tag{6.134}$$

We make the hypothesis

$$\boldsymbol{\xi}_n = \beta^n \boldsymbol{\xi}_0 . \tag{6.135}$$

Inserting it into (6.126) yields

$$\Big(\beta^{n+1} - e^{-\gamma\Delta}\beta^n - e^{-\gamma(\Delta-\tau_2')}\tilde{A}_2\beta^{n-M_2} \\ - e^{\gamma\tau_1'}\tilde{A}_1\beta^{n+1-M_1}\Big)\boldsymbol{\xi}_0 = 0 , \tag{6.136}$$

or equivalently

$$\left(\beta - e^{-\gamma\Delta} - \beta^{-M_2}e^{-\gamma(\Delta-\tau_2')}\tilde{A}_2 - e^{\gamma\tau_1'}\tilde{A}_1\beta^{1-M_1}\right)\boldsymbol{\xi}_0 = 0 . \tag{6.137}$$

In order to discuss the solution of this equation, we consider a few examples. (1) $M_1 = 1, \quad M_2 = 0$. In this case, (6.137) reduces to

$$\left(\beta - e^{-\gamma\Delta} - e^{-\gamma(\Delta-\tau_2')}\tilde{A}_2 - e^{\gamma\tau_1'}\tilde{A}_1\right)\boldsymbol{\xi}_0 = 0 . \tag{6.138}$$

We first study the following eigenvalue equation

$$\left(e^{-\gamma(\Delta-\tau_2')}\tilde{A}_2 + e^{\gamma\tau_1'}\tilde{A}_1\right)\mathbf{v}_\mu = \lambda_\mu\mathbf{v}_\mu, \quad \mu = 1, ..., N . \tag{6.139}$$

Using its solution in (6.138) with $\boldsymbol{\xi}_0 = \mathbf{v}_\mu$, we obtain the corresponding eigenvalues

$$\beta_\mu = e^{-\gamma\Delta} + \lambda_\mu(\Delta, \tau_1', \tau_2') . \tag{6.140}$$

(2) Here we consider the case $M_2 = M, M_1 = M + 1$. In this case we have to solve the eigenvalue equation

$$\left(\beta^{M+1} - e^{-\gamma\Delta}\beta^M - e^{-\gamma(\Delta-\tau_2')}\tilde{A}_2 - e^{\gamma\tau_1'}\tilde{A}_1\right)\boldsymbol{\xi}_0 = 0\,, \tag{6.141}$$

which because of (6.139) can be transformed into

$$\beta^{M+1} - e^{-\gamma\Delta}\beta^M - \lambda_\mu = 0\,. \tag{6.142}$$

Using the results of Sect. 5.7, we can easily solve this equation in the case that λ_μ is a small quantity.

There are two kinds of solutions, namely

$$\beta_\mu = e^{-\gamma\Delta} + \lambda_\mu e^{M\Delta\gamma}\,, \quad \mu = 1, ..., N \tag{6.143}$$

and, for $M \geq 1$,

$$\beta_{j,\mu} = \left(-\lambda_\mu e^{\gamma\Delta}\right)^{1/M} e^{2\pi ij/M}\,, \quad \begin{array}{l} j = 0, ..., M-1 \\ \mu = 1, ..., N\,. \end{array} \tag{6.144}$$

Since β determines the damping of the deviations $\boldsymbol{\xi}$ from the phase-locked state ϕ, and also the instability of ϕ (see below), the dependence of β on the delay time $\tau = M\Delta$ is of particular interest. According to (6.143), for $\lambda_\mu > 0$ the damping becomes weaker when the delay time $\tau = M\Delta$ increases. Since for a stable solution $|\lambda_\mu e^{\gamma\Delta}| < 1$, the M-dependence of β indicates the same trend even more. In other words: the larger the time delay, the less stable the phase-locked solution is; it may even become unstable while in cases without time delays it is still stable. It is also noteworthy that according to (6.144) the solutions are oscillatory, because β is complex. We are now in a position to perform the complete solution to (6.126) taking into account the initial conditions (6.131)–(6.133). To this end we realize that the complete solution is given by

$$\boldsymbol{\xi}_n = \sum_\mu \left(c_\mu \beta_\mu^n + \sum_{j=0}^{M-1} c_{j,\mu}\beta_{j,\mu}^n\right)\mathbf{v}_\mu\,. \tag{6.145}$$

To calculate the coefficients $c_\mu, c_{j,\mu}$, we have to insert this expression into (6.126) (with $\hat{B} = 0$) for $n = 0, 1, ..., M = \max(M_1 - 1, M_2)$, which yields equations that have been exemplified by (6.131)–(6.133). This yields $M \times N$ equations for the $M \times N$ unknown coefficients $c_\mu, c_{j,\mu}$. The explicit solution depends, of course, on the explicit form of $\tilde{A}_1, \tilde{A}_2$.

Let us finally consider the general case (6.137), but assume

$$M_2 \geq M_1\,. \tag{6.146}$$

The case in which $M_2 \leq M_1 - 1$ can be treated similarly. We then cast (6.137) into the form

$$\left(\beta^{M_2+1} - e^{-\gamma\Delta}\beta^{M_2} - e^{-\gamma(\Delta-\tau_2')}\tilde{A}_2 - e^{\gamma\tau_1'}\tilde{A}_1\beta^{M_2+1-M_1}\right)\boldsymbol{\xi}_0 = 0\,. \quad (6.147)$$

This relationship can be simplified if we assume that the matrices $\tilde{A}_1$ and $\tilde{A}_2$ possess the same eigenvectors $\mathbf{v}_\mu$ but possibly different eigenvalues $\lambda_{1,\mu}, \lambda_{2,\mu}$. From (6.147) we then derive

$$\beta^{M_2+1} + e^{-\gamma\Delta}\beta^{M_2} - d_1\lambda_{1,\mu}\beta^{M_2+1-M_1} - d_2\lambda_{2,\mu} = 0\,, \quad (6.148)$$

where $d_1 = e^{-\gamma\tau_1'}$, $d_2 = e^{-\gamma(\Delta-\tau_2')}$. The discussion of the corresponding eigenvalue equation for β must be left to a further analysis. Otherwise we can proceed as before.

6.8 Many Different Delay Times*

For the interested reader and for the sake of completeness, we quote the extension of the results of the preceding section to many different delay times. We put

$$\tau_\ell = M_\ell\Delta + \tau_\ell', \quad 0 \leq \tau_\ell' < \Delta, \; M \text{ an integer and } \ell = 1, 2, ...L\,. \quad (6.149)$$

The relation (6.103) generalizes to

$$\Delta = \frac{1}{c}\left(2\pi - (1/\gamma)\sum_{\ell=1}^{L} A_\ell\right) \quad (6.150)$$

and (note (6.128)!) the recursive relations (6.126) become

$$\boldsymbol{\xi}_{n+1} = e^{-\gamma\Delta}\boldsymbol{\xi}_n + e^{-\gamma\Delta}\sum_{\ell=1}^{L}\tilde{A}_\ell e^{\gamma\tau_\ell'}\boldsymbol{\xi}_{n-M_\ell} + e^{\gamma\tau_1'}\hat{B}\left((n+1)\Delta + \tau_1'; n\Delta + \tau_1'\right)\,. \quad (6.151)$$

The eigenvalue equations for β become particularly simple if

$$M_\ell = M \quad (6.152)$$

so that (6.141) generalizes to

$$\left(\beta^{M+1} - e^{-\gamma\Delta}\beta^M - e^{-\gamma\Delta}\sum_{\ell=1}^{L} e^{\gamma\tau_\ell'}\tilde{A}_\ell\right)\boldsymbol{\xi}_0 = 0\,. \quad (6.153)$$

The further steps can be performed as before.

6.9 Phase Waves in a Two-Dimensional Neural Sheet

So far the index j as well as other indices simply served for an enumeration of the neurons. In this section we want to use them to indicate the positions of the neurons. To this end we consider a two-dimensional neural sheet in which the neurons, at least approximately, occupy the sites of a lattice with the axes x and y. These axes need not be orthogonal to each other. Along each axis, the distance between neighboring sites be a. We replace the index j by an index vector,

$$j \to \mathbf{j} = (j_x, j_y) , \tag{6.154}$$

where $j_x = an_x, \quad j_y = an_y$, and n_x, n_y are integer numbers $1, 2, ..., L_x$ (or L_y). Thus the extensions of the neural sheet in its two directions are $J_x = aL_x$ and $J_y = aL_y$. Since the coefficients $A_{j,\ell}$ or $a_{j,\ell}$ contain two indices, we replace the latter by two vectors $\mathbf{j}, \boldsymbol{\ell}$. We consider an important case in which the coefficients $a_{j,\ell} \equiv a_{\mathbf{j},\boldsymbol{\ell}}$ depend on the difference of the vectors $\mathbf{j}$ and $\boldsymbol{\ell}$, for instance on the distance between the neural sites. In other words, we assume

$$a_{\mathbf{j},\boldsymbol{\ell}} = a(\boldsymbol{\ell} - \mathbf{j}) .$$

To make things mathematically simple, we assume that $a(\mathbf{j})$ is periodic with periods J_x and J_y in the corresponding directions. We are now in a position to solve (6.139) in a very simple fashion. Using the vectors $\mathbf{j}$ and $\boldsymbol{\ell}$, we write (6.139) in the form

$$\sum_{\boldsymbol{\ell}} a(\boldsymbol{\ell} - \mathbf{j}) v(\boldsymbol{\ell}) = \lambda v(\boldsymbol{\ell}) , \tag{6.155}$$

where the vector $\mathbf{v}$ of (6.139) has the components $v(\boldsymbol{\ell})$. To solve (6.155), we make the hypothesis

$$v(\boldsymbol{\ell}) = N e^{i\boldsymbol{\kappa}\boldsymbol{\ell}} , \tag{6.156}$$

where N is a normalization constant and $\boldsymbol{\kappa}$ is a wave vector with components κ_x, κ_y. Inserting (6.156) in (6.155) and multiplying both sides by $e^{-i\boldsymbol{\kappa}\boldsymbol{\ell}}$ yields

$$\sum_{\boldsymbol{\ell}} a(\boldsymbol{\ell} - \mathbf{j}) e^{i\boldsymbol{\kappa}(\mathbf{j}-\boldsymbol{\ell})} = \lambda_{\boldsymbol{\kappa}} , \tag{6.157}$$

where we equipped λ with the index $\boldsymbol{\kappa}$, because the l.h.s. certainly depends on $\boldsymbol{\kappa}$. According to (6.155), the eigenvalues $\lambda_{\boldsymbol{\kappa}}$ are the Fourier transform of $a(\boldsymbol{\ell})$, whereby we use the new summation index $\boldsymbol{\ell}$ instead of $\boldsymbol{\ell} - \mathbf{j}$. Clearly the vectors $\mathbf{v} = (v(\boldsymbol{\ell}))$ depend also on $\boldsymbol{\kappa}$ so that we write

$$\mathbf{v} = \mathbf{v}_{\boldsymbol{\kappa}} . \tag{6.158}$$

We can now interpret (6.145) in more detail. The displacements $\boldsymbol{\xi}$ of the phases are a superposition of *spatial* waves. To bring out the *time dependence* of the amplitudes of these waves more explicitly, we write the coefficients of (6.145) in the form

$$\beta^n = e^{(i\omega-\gamma)t_n}, \tag{6.159}$$

where t_n are the discrete times. Since β can become complex in the case of delays, the frequencies ω become $\neq 0$ in this case. Otherwise waves are purely damped in time. The occurrence of excitation waves in brain tissue is known both in vivo and in vitro, but further detailed studies are needed.

6.10 Stability Limits of Phase-Locked State

In Sect. 6.8 we established a general relation for the pulse interval Δ, namely (6.150). Since this interval, because of its definition, cannot become negative, something particular must happen in the limit $\Delta \to 0$. We may suspect that this has something to do with the stability of the phase-locked state. Since (6.153) allows us to study the stability limit quite generally, we want to establish a relation between (6.153) and (6.150). To present this relation as simply as possible, we consider the case of no delay, i.e. (6.138) with $\tilde{A}_2 = 0, \tau_1' = 0$. Because of (6.86) and (6.115), among the eigenvalues λ_μ of (6.139) is the following one

$$\lambda = \dot{\phi}^{-1} A \equiv \dot{\phi}^{-1} \sum_k A_{jk,1}\,. \tag{6.160}$$

It belongs to a vector $\mathbf{v}$ with equal components. According to (6.98), we may make the estimate $\dot{\phi} = 2\pi/\Delta$, so that

$$\lambda = \Delta A/2\pi\,. \tag{6.161}$$

Writing $\beta_\mu \equiv \beta$ in the form $\exp(\Lambda\Delta)$, we note that the *Lyapunov exponent* Λ is given by

$$\Lambda = \frac{1}{\Delta} \ln \beta\,. \tag{6.162}$$

From (6.140) and (6.161) we obtain for $\Delta \to 0$

$$\Lambda = \frac{1}{\Delta} \ln\left(e^{-\gamma\Delta} + \Delta A/2\pi\right) \approx \frac{1}{\Delta} \ln\left(1 + \Delta(A/2\pi - \gamma)\right) \approx A/2\pi - \gamma\,. \tag{6.163}$$

The vanishing of the Lyapunov exponent, $\Lambda = 0$, indicates instability. This condition coincides with the vanishing of Δ (6.150). This coincidence of $\Delta = 0$ with the instability limit holds, however, only, if the real parts of all other eigenvalues λ_μ are not larger than λ, otherwise the instability occurs earlier.

We assumed, of course, $\lambda > 0, A_1 > 0$. While the "λ-instability" retains the phase-locked state but leads to an exponential increase of the pulse rate (see Sect. 6.12 below), the other "λ_μ-instabilities" lead to a destruction of the phase-locked state, giving rise to spatiotemporal activity patterns of the neurons. This can be seen as follows: The eigenvalue λ is connected with a space-independent eigenvector $\mathbf{v}$, whereas the other eigenvalues belong to space-dependent eigenvectors, e.g. to $\exp(i\boldsymbol{\kappa}\mathbf{n})$ (or their corresponding real or imaginary parts). These patterns may become stabilized if the saturation of S (6.3) is taken into account.

6.11 Phase Noise*

After having explored the impact of delay on the stability of the phase-locked state, we turn to the study of the impact of noise. In order not to overload our presentation, we treat the example of a *single* delay time τ. It is hoped that in this way the general procedure can be demonstrated. We assume that a phase-locked state with $\phi_j = \phi$ exists and we study small deviations $\boldsymbol{\xi} = (\xi_1, ..., \xi_N)$ from that state, i.e. we put

$$\phi_j = \phi + \xi_j \,. \tag{6.164}$$

In a generalization of Sect. 6.7, for small noise and corresponding small deviations ξ_j, the following difference equations for $\boldsymbol{\xi} = \boldsymbol{\xi}_n$ at times $t_n = n\Delta$, where Δ is the pulse interval, can be derived

$$\boldsymbol{\xi}_{n+1} = e^{-\gamma\Delta}\boldsymbol{\xi}_n + \tilde{A}\boldsymbol{\xi}_{n-M} + \hat{\mathbf{B}}_n \,. \tag{6.165}$$

$\tilde{A}$ is a time-independent matrix proportional to $A = (A_{jk})$, and

$$\hat{B}_{\ell,n} = \int_{t_{n-1}}^{t_n} e^{-\gamma(t_n-\sigma)}\hat{F}_\ell(\sigma)d\sigma \,. \tag{6.166}$$

We measure the delay time τ in units of Δ, i.e. $\tau = M\Delta$, M an integer. The components $\hat{B}_{\ell,n}$ of the fluctuating forces are assumed to obey

$$< \hat{B}_{\ell,n}\hat{B}_{\ell',n'} >= Q_{\ell\ell'}\delta_{nn'} \,. \tag{6.167}$$

We first seek the eigenvectors and eigenvalues of the matrix $\tilde{A}$ (assuming that Jordan's normal form has only diagonal elements)

$$\tilde{A}\mathbf{v}_\mu = \lambda_\mu\mathbf{v}_\mu \,. \tag{6.168}$$

We decompose $\boldsymbol{\xi}_n$ according to

$$\boldsymbol{\xi}_n = \sum_\mu \xi_n^{(\mu)}\mathbf{v}_\mu \,. \tag{6.169}$$

We project both sides of (6.165) on the eigenvectors $\mathbf{v}_\mu$ and obtain

$$\xi_{n+1}^{(\mu)} = e^{-\gamma\Delta}\xi_n^{(\mu)} + \lambda_\mu \xi_{n-M}^{(\mu)} + \hat{B}_n^{(\mu)}, \tag{6.170}$$

where $\hat{B}_n^{(\mu)} = \left(\mathbf{v}_\mu^+ \hat{\mathbf{B}}_n\right)$. By means of a correlation matrix K we can express correlations between the components of $\boldsymbol{\xi}_N, \boldsymbol{\xi}_{N'}$ by means of correlations between $\xi_N^{(\mu)}$ and $\xi_{N'}^{(\mu')}$

$$< (\boldsymbol{\xi}_N K \boldsymbol{\xi}_{N'}) > = \sum_{\mu\mu'} \left(\mathbf{v}_\mu K \mathbf{v}_{\mu'}\right) \; < \xi_N^{(\mu)} \xi_{N'}^{(\mu')} > . \tag{6.171}$$

We evaluate the correlation functions on the r.h.s. of (6.171) and start with $\mu = \mu'$. In order not to overload the formulas with indices, we drop the suffix μ everywhere in (6.170) and put

$$\lambda_\mu = a \, . \tag{6.172}$$

We first study

$$\xi_{n+1} = e^{-\gamma\Delta}\xi_n + a\xi_{n-M} + \delta_{nn_0} \tag{6.173}$$

with the initial conditions

$$\xi_n = 0 \quad \text{for} \quad n \leq n_0 \tag{6.174}$$

and

$$\xi_{n_0+1} = 1 \, . \tag{6.175}$$

The general solution of

$$\xi_{n+1} = e^{-\gamma\Delta}\xi_n + a\xi_{n-M}, \quad n > n_0 \, , \tag{6.176}$$

can be written in the form

$$\xi_n = c_1\beta_1^n + c_2\beta_2^n ... + c_{M+1}\beta_{M+1}^n \, , \tag{6.177}$$

where β_j are the solutions of the eigenvalue equation

$$\beta^{M+1} - e^{-\gamma\Delta}\beta^M - a = 0 \, . \tag{6.178}$$

Taking into account the initial conditions (6.174) and (6.175), we obtain the following equations

$$n = n_0 : \xi_{n_0+1} = 1 \, , \tag{6.179}$$

$$n = n_0 + 1 : \xi_{n_0+2} = e^{-\gamma\Delta}\xi_{n_0+1} + a\xi_{n_0+1-M} = e^{-\gamma\Delta} \, , \tag{6.180}$$

$$n = n_0 + 2 : \xi_{n_0+3} = e^{-\gamma\Delta}\xi_{n_0+2} = e^{-2\gamma\Delta} \, , \tag{6.181}$$

$$\vdots$$

$$n = n_0 + M : \xi_{n_0+1+M} = e^{-\gamma\Delta M}\,, \tag{6.182}$$

$$n = n_0 + M + 1 : \xi_{n_0+2+M} = e^{-\gamma\Delta(M+1)} + a\,. \tag{6.183}$$

Inserting (6.177) into (6.179)–(6.183), we obtain the following equations

$$c_1\beta_1^{n_0+1} + c_2\beta_2^{n_0+1} + ... + c_{M+1}\beta_{M+1}^{n_0+1} = 1\,, \tag{6.184}$$

$$c_1\beta_1^{n_0+2} + c_2\beta_2^{n_0+2} + ... + c_{M+1}\beta_{M+1}^{n_0+2} = e^{-\gamma\Delta}\,, \tag{6.185}$$

$$\vdots$$

$$c_1\beta_1^{n_0+M+2} + c_2\beta_2^{n_0+M+2} + ... + c_{M+1}\beta_{M+1}^{n_0+M+2} = e^{-(M+1)\gamma\Delta} + a\,. \tag{6.186}$$

The solutions c_j of these equations can be simplified by using the hypothesis

$$c_j\beta_j^{n_0+1} = d_j\,, \tag{6.187}$$

which converts (6.184)–(6.186) into

$$d_1 + d_2 + ... + d_{M+1} = 1\,, \tag{6.188}$$

$$d_1\beta_1 + d_2\beta_2 + ... + d_{M+1}\beta_{M+1} = e^{-\gamma\Delta}\,, \tag{6.189}$$

$$\vdots$$

$$d_1\beta_1^M + d_2\beta_2^M ... + d_{M+1}\beta_{M+1}^M = e^{-M\gamma\Delta}\,, \tag{6.190}$$

$$d_1\beta_1^{M+1} + d_2\beta_2^{M+1} ... + d_{M+1}\beta_{M+1}^{M+1} = e^{-(M+1)\gamma\Delta} + a\,. \tag{6.191}$$

The original hypothesis (6.177) can thus be written in the form

$$\xi_{n,n_0} = d_1\beta_1^{n-(n_0+1)} + ... + d_{M+1}\beta_{M+1}^{n-(n_0+1)}\,, \tag{6.192}$$

where, in order to make the dependence of ξ_n on n_0 explicit, we used the substitution

$$\xi_n \to \xi_{n,n_0}\,. \tag{6.193}$$

We are now in a position to deal with (6.170) that we recast into the form

$$\tilde{\xi}_{n+1} = e^{-\gamma\Delta}\tilde{\xi}_n + a\tilde{\xi}_{n-M} + \sum_{n_0}\delta_{nn_0}\hat{B}_{n_0}\,. \tag{6.194}$$

Since (6.194) is a linear equation, its solution can be found by means of the solutions of (6.173)

$$\tilde{\xi}_n = \sum_{n_0=0}^{n-1}\xi_{n,n_0}\hat{B}_{n_0}\,. \tag{6.195}$$

In the following we will use the index n' instead of n_0. Inserting (6.192) into (6.195), we thus obtain

$$\xi_n = \sum_{j=1}^{M+1} d_j \sum_{n'=0}^{n-1} \beta_j^{n-(n'+1)} \hat{B}_{n'} \,. \tag{6.196}$$

Having in mind the same index μ, we may now evaluate the correlation function occurring in (6.171), namely

$$\begin{aligned} < \xi_N \xi_{N'} > \; &= \sum_{j=1}^{M+1} \sum_{j'=1}^{M+1} d_j d'_j \sum_{n=0}^{N-1} \beta_j^{N-(n+1)} \\ &\times \sum_{n'=0}^{N'-1} \beta_{j'}^{N'-(n'+1)} < \hat{B}_n \hat{B}_{n'} > \,. \end{aligned} \tag{6.197}$$

Assuming $N \geq N'$ and using

$$< \hat{B}_n \hat{B}_{n'} > = Q \delta_{nn'} \tag{6.198}$$

as well as (6.167), we readily obtain

$$< \xi_N \xi_{N'} > = \sum_{j,j'=1}^{M+1} d_j d_{j'} \sum_{n=0}^{N'-1} \beta_j^{N-(n+1)} \beta_{j'}^{N'-(n+1)} Q \,. \tag{6.199}$$

The sum over n can readily be performed so that we obtain

$$< \xi_N \xi_{N'} > = Q \sum_{j,j'=1}^{M+1} d_j d_{j'} \beta_j^{N-N'} \left(1 - \beta_j \beta_{j'}\right)^{-1} \,. \tag{6.200}$$

Under the assumption

$$\mid \beta_j \beta_{j'} \mid \ll 1 \tag{6.201}$$

and using (6.188), we obtain the very concise form

$$< \xi_N \xi_{N'} > = Q \sum_{j=1}^{M+1} d_j \beta_j^{N-N'} \,. \tag{6.202}$$

Provided

$$0 \leq N - N' \leq M + 1 \,, \tag{6.203}$$

the expression (6.202) can be considerably simplified by means of the former relations (6.188)–(6.191). Thus we obtain, for instance

$$< \xi_N \xi_N > = Q \tag{6.204}$$

and

$$< \xi_{N+1}\xi_N > = Qe^{-\gamma\Delta} . \tag{6.205}$$

By means of the substitutions

$$\beta_j \to \beta_{j,\mu}, \quad d_j \to d_{j,\mu} , \tag{6.206}$$

$$\xi_N \to \xi_N^{(\mu)} , \tag{6.207}$$

$$B_n \to B_n^{(\mu)} , \tag{6.208}$$

$$Q^{(\mu,\mu')} = < \left(\mathbf{v}_\mu^+ \hat{B}_n\right) \left(\mathbf{v}_{\mu'}^+ \hat{B}_n\right) > , \tag{6.209}$$

we can generalize the result (6.202) to

$$< \xi_N^{(\mu)} \xi_{N'}^{(\mu')} > = Q^{(\mu,\mu')} \sum_{j=1}^{M+1} d_{j,\mu} \beta_{j,\mu}^{N-N'}, \quad N \geq N' , \tag{6.210}$$

which presents under the condition (6.201) our central result. As we can see, in the case of a delay with the delay time $\tau = M\Delta$, where Δ is the pulse interval of the phase-locked state, the correlation function is not only determined by means of the strengths $Q^{(\mu,\mu')}$ of the fluctuating forces, but also by the eigenvalues of (6.178). Since the eigenvalues may be complex, the correlation function (6.210) may show oscillatory behavior as a function of $N - N'$. It is interesting that the index μ' drops out provided (6.201) is assumed. The case in which $N' > N$ can be covered by exchanging the indices μ and μ'.

Exercises. (1) Why does ξ_{n,n_0} play the role of a Green's function for a set of difference equations?
(2) Derive $< \xi_N^{(\mu)} \xi_{N'}^{(\mu')} >$ for arbitrary β values, i.e. without the assumption (6.201).
(3) Denote the vector components of $\boldsymbol{\xi}_n$ by $\xi_{n,\ell}$. Derive a formula for $< \xi_{N,\ell}\xi_{n',\ell} >$.
Hint: Use (6.171) with suitable K.
(4) Determine (6.210) for $M = 1$ explicitly.

6.12 Strong Coupling Limit. The Nonsteady Phase-Locked State of Many Neurons

Our solution of the basic (6.1) and (6.3) of the neural net relies on the assumption that the system operates in the linear range of S. This requires a not too strong coupling as expressed by the coupling coefficients $A_{j\ell;m}$ (synaptic strengths) or a not too small damping constant γ. We also noted that in the linear range of S the phase-locked state is only stable if the just-mentioned conditions are satisfied.

In this section and the following we want to study what happens if these conditions are violated. To this end we first study the case of phase-locking in the *linear range* of S. While for weak enough coupling we obtained a *steady* phase-locked state, for strong coupling a new kind of behavior will occur, namely pulse-shortening until the nonlinearity of S leads to a shortest pulse-interval. This nonlinear case will be treated in the following Sect. 6.13. In that section, our goal will be to determine the pulse interval Δ as a function of the neural parameters. Hereby we treat the *phase-locked* case. In order not to overload our presentation, we neglect time delays and fluctuations.

Our starting point is Sect. 6.2, where according to (6.17) and (6.18) the phase-locking equation is given by

$$\ddot{\phi} + \gamma\dot{\phi} = Af(\phi) + C, \quad A > 0. \tag{6.211}$$

We make the substitution

$$\dot{\phi} = C/\gamma + x(t) \tag{6.212}$$

that converts (6.211) into

$$\dot{x} + \gamma x = Af(\phi). \tag{6.213}$$

The solution of (6.213) proceeds as in Sect. 5.3, whereby we find the recursion relation for $x_n \equiv x(t_n)$ (see (5.40))

$$x_{n+1} = e^{-\gamma\Delta_n} x_n + A, \tag{6.214}$$

where we assume, however, that the pulse intervals $\Delta_n \equiv t_{n+1} - t_n$ depend on n in order to cover non-stationary cases. In an extension of Sect. 5.3, we find the following equation for Δ_n (with $c = C/\gamma$) (see (5.46))

$$c\Delta_n + \frac{1}{\gamma} x_n \left(1 - e^{-\gamma\Delta_n}\right) = 2\pi. \tag{6.215}$$

The coupled equations (6.214) and (6.215) for the dependence of the variables x_n and Δ_n on n can be easily solved numerically. For our present purpose it is sufficient to discuss the solution qualitatively. As we will see, Δ_n decreases as n increases. Therefore, for a sufficiently large n we may assume that $\gamma\Delta_n \ll 1$, so that

$$\left(1 - e^{-\gamma\Delta_n}\right) \approx \gamma\Delta_n. \tag{6.216}$$

Equations (6.214) and (6.215) then simplify to

$$x_{n+1} - x_n = -\gamma\Delta_n x_n + A \tag{6.217}$$

and

$$c\Delta_n + x_n\Delta_n = 2\pi\,, \tag{6.218}$$

respectively. Resolving (6.218) for Δ_n yields

$$\Delta_n = \frac{2\pi}{x_n + c}\,, \tag{6.219}$$

which can be inserted into (6.217) to yield

$$x_{n+1} - x_n = A - 2\pi\gamma\frac{x_n}{x_n + c}\,. \tag{6.220}$$

Figure 6.5 shows a plot of the r.h.s. of (6.220), i.e. of

$$R(x_n) = A - 2\pi\gamma\frac{x_n}{x_n + c}, \quad A > 0 \tag{6.221}$$

versus x_n. Figure 6.5a refers to the case of small coupling A,

$$A - 2\pi\gamma < 0\,. \tag{6.222}$$

In this case $R(x_n)$ becomes zero at a specific value of $x_n = x_s$, and the solution x_n of (6.220) becomes independent of n, i.e. of time. In other words, x_s is a fixed point of (6.220). Thus we recover the solution obtained in Sect. 5.3. Because of (6.219), the pulse intervals become equidistant.

Figure 6.5b refers to the case of strong coupling

$$A - 2\pi\gamma > 0\,. \tag{6.223}$$

Because $R(x_n)$ never crosses the abscissa, according to (6.220) $x_{n+1} - x_n$ is always positive, i.e. x_n increases forever. For large enough x_n, (6.220) simplifies to

$$x_{n+1} - x_n = A - 2\pi\gamma\,, \tag{6.224}$$

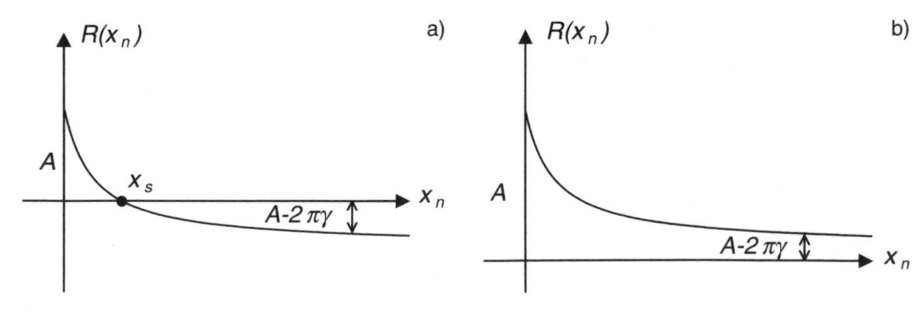

Fig. 6.5. a) The function $R(x_n)$ (6.221) versus x_n for $A - 2\pi\gamma < 0$; **b)** Same as a), but $A - 2\pi\gamma > 0$

which possesses the solution

$$x_n = (A - 2\pi\gamma)n + \text{const.} \tag{6.225}$$

as we may immediately verify. Bearing in mind that x_n contributes to the pulse rate $\dot{\phi}$ according to (6.212), we obtain our final result that $\dot{\phi}$ increases linearly with n. Equivalently, the pulse intervals Δ_n decrease according to (6.219) and (6.225)

$$\Delta_n = \frac{2\pi}{(A - 2\pi\gamma)n + \text{const.}} . \tag{6.226}$$

Because of the definition of $\Delta_n \equiv t_{n+1} - t_n$, we may now establish a relation between t_n and n. As a simple analysis (see the exercise at the end of this section) reveals, t_n increases logarithmically with n, i.e.

$$t_n \approx \frac{2\pi}{(A - 2\pi\gamma)} \ln n + \text{const.} \tag{6.227}$$

for n large. On the other hand, we may express n by t_n to obtain

$$n = n_0 \exp\left[(A/2\pi - \gamma)t_n\right] . \tag{6.228}$$

This means that Δ_n *decreases* exponentially with time, or that x_n and thus the pulse rate $\dot{\phi}$ *increases* exponentially. Let us now recall that the pulse rate is directly given by the saturation function S according to (6.3), i.e. by

$$\dot{\phi} = S . \tag{6.229}$$

If $\dot{\phi}$ becomes large, the quasilinear range of S is abandoned and S becomes a constant, i.e. it saturates,

$$S = S_{\max} . \tag{6.230}$$

Summing up the results of this section, we may state that in the strong coupling case the pulse rate increases exponentially with time until it saturates. Our derivation of the time dependence of Δ_n and x_n has been systematic, but a bit clumsy. A more direct way, based on the guess that Δ_n decays exponentially is this: In order to solve (6.215) we make the hypothesis

$$\Delta_n = \Delta_0 e^{-\Gamma t_n} \tag{6.231}$$

with yet unknown Δ_0 and Γ. Inserting this into (6.215) and neglecting the small term $c\Delta_n$ in (6.218) yields

$$x_n \Delta_0 e^{-\Gamma t_n} = 2\pi . \tag{6.232}$$

We then make the hypothesis

$$x_n = x_0 e^{\Gamma t_n} , \tag{6.233}$$

from which follows

$$x_0 \Delta_0 = 2\pi \,. \tag{6.234}$$

Inserting (6.233) into (6.214) yields

$$x_0 e^{\Gamma t_{n+1}} = e^{-\gamma \Delta_n} x_0 e^{\Gamma t_n} + A \,. \tag{6.235}$$

Using

$$t_{n+1} = t_n + \Delta_n \,, \tag{6.236}$$

we obtain

$$x_0 e^{\Gamma \Delta_n} = e^{-\gamma \Delta_n} x_0 + A e^{-\Gamma t_n} \,. \tag{6.237}$$

Using the hypothesis (6.231) on both sides of (6.237) and expanding the exponential functions, we obtain

$$x_0(1 + \Delta_0 \Gamma e^{-\gamma t_n}) = x_0(1 - \gamma \Delta_0 e^{-\Gamma t_n}) + A e^{-\Gamma t_n} \,. \tag{6.238}$$

Comparing the corresponding terms on the left- and right-hand side of (6.238), we obtain

$$x_0 \Delta_0 \Gamma = -\gamma x_0 \Delta_0 + A \,, \tag{6.239}$$

from which we conclude, with the help of (6.234),

$$\Gamma = -\gamma + A/(2\pi) \,. \tag{6.240}$$

This agrees with the results obtained using our "systematic" procedure above.

Exercise. (1) Prove (6.227)
Hint: Approximate $\Delta_n \equiv t_{n+1} - t_n$ by dt/dn. Solve the resulting differential equation for large n.

6.13 Fully Nonlinear Treatment of the Phase-Locked State*

Our approach has been nonlinear in so far as the phase angles ϕ appear under a series of δ-functions. On the other hand, we linearized the sigmoid function S. In this section we want to perform a fully nonlinear treatment in which also the nonlinearity of S is maintained. In contrast to Sect. 6.12, where we studied the transient phase-locked solution, here we deal with the steady-state solution. Since the most important observable is the pulse interval Δ, we will determine it here. The calculations are somewhat lengthy. Thus the speedy reader may skip all the intermediate steps and proceed to the final result that relates Δ to the neural parameters, such as synaptic strengths, etc. These results are presented in (6.270) with its special cases for linear S (6.271) and saturated S (6.276).

First we remind the reader of the basic equations for the dendritic currents

$$\dot{\psi}_m = -\gamma\psi_m + \sum_k a_{mk} f(\phi_k) \tag{6.241}$$

and the axonal phases

$$\dot{\phi}_k = S\left(\sum_m c_{km}\psi_m(t-\tau'_{km}) + p_{\mathrm{ext,k}}, \Theta_k\right). \tag{6.242}$$

The notation is the same as in our earlier chapters. By use of the Green's function of Sect. 4.4, we can express ψ_m directly by the last term in (6.241) and insert the result into (6.242). In anticipation of the phase-locked state, we put $\phi_k = \phi$ and obtain

$$\dot{\phi}(t) = S\left(A_k \int_0^t e^{-\gamma(t-\sigma)} f(\phi(\sigma))d\sigma + p_{\mathrm{ext,k}}(t), Q_k\right), \tag{6.243}$$

where

$$A_k = \sum_{k'}\sum_m c_{km} a_{mk'} . \tag{6.244}$$

Since the l.h.s. is independent of the index k, so must be the r.h.s. A sufficient condition will be that $A_k, p_{\mathrm{ext,k}}$ and Q_k are independent of k. Using the explicit form of f and assuming a steady state, we first evaluate the integral of (6.243), i.e.

$$\int_0^t e^{-\gamma(t-\sigma)} \sum_{n=0}^{N} \delta(\sigma - n\Delta)d\sigma . \tag{6.245}$$

We use the definition

$$N = [t/\Delta], \tag{6.246}$$

where the square bracket means: Take the biggest integer number that is smaller than or equal to t/Δ. Using the property of the δ-function and summing the resulting geometric series, we obtain in the limit of large times t

$$(6.245) = e^{-\gamma t}\sum_{n=0}^{N} e^{\gamma n\Delta} = e^{\gamma((N+1)\Delta - t)}\frac{1}{e^{\gamma\Delta}-1}. \tag{6.247}$$

We use the Naka–Rushton formula for S

$$S(X) = \frac{rX^M}{Q^M + X^M}, \tag{6.248}$$

where in the present case

$$X = Ae^{\gamma((N+1)\Delta - t)} \frac{1}{e^{\gamma\Delta} - 1} + p \tag{6.249}$$

and

$$p_{\text{ext,k}}(t) = p \,. \tag{6.250}$$

Using (6.245)–(6.250) in (6.243), we can immediately determine the phase velocity $\dot{\phi}$, provided we know the pulse interval Δ. In order to fix the still unknown Δ, we postulate

$$\int\limits_{t_N+\epsilon}^{t_{N+1}+\epsilon} \dot{\phi}(\sigma) d\sigma = \int\limits_{t_N+\epsilon}^{t_{N+1}-\epsilon} \dot{\phi}(\sigma) d\sigma = 2\pi \,. \tag{6.251}$$

In the following, we will evaluate the integral explicitly for $M = 1$ and show in the exercise how it can be determined for integer $M > 1$.

We write (6.248) in the form

$$S(X) = \frac{r(X/Q)^M}{1 + (X/Q)^M} \tag{6.252}$$

and introduce the abbreviation

$$X/Q = \tilde{A} \cdot e^{\gamma(t_{N+1} - t)} + \tilde{p} \,, \tag{6.253}$$

where

$$\tilde{A} = \frac{A}{Q} \left(e^{\gamma\Delta} - 1\right)^{-1} , \frac{p}{Q} = \tilde{p} \,. \tag{6.254}$$

After shifting the limits of the integration, we have to evaluate

$$\int\limits_0^\Delta \frac{\left(\tilde{A}e^{\gamma\sigma} + \tilde{p}\right)^M}{1 + \left(\tilde{A}e^{\gamma\sigma} + \tilde{p}\right)^M} d\sigma \,. \tag{6.255}$$

To this end we put

$$\tilde{A}e^{\gamma\sigma} + \tilde{p} = y \tag{6.256}$$

and accordingly

$$\tilde{A}\gamma e^{\gamma\sigma} d\sigma = dy \,. \tag{6.257}$$

From (6.256) and (6.257) we deduce

$$d\sigma = \frac{1}{\gamma} \frac{1}{y - \tilde{p}} dy \,, \tag{6.258}$$

so that we are, eventually, left with the evaluation of the integral

$$\frac{1}{\gamma}\int_{y_o}^{y_1}\frac{y^M}{1+y^M}\frac{1}{(y-\tilde{p})}dy\,. \tag{6.259}$$

We will determine the integration limits below. In order to perform the integration, we use the decomposition

$$\frac{y^M}{1+y^M}\frac{1}{y-\tilde{p}}=\frac{g(y)}{1+y^M}+\frac{\alpha}{y-\tilde{p}}\,, \tag{6.260}$$

where $g(y)$ is determined by

$$g(y)(y-\tilde{p})=y^M(1-\alpha)-\alpha\,. \tag{6.261}$$

The unknown coefficient α can be fixed by putting $y=\tilde{p}$ on both sides, which yields

$$0=\tilde{p}^M(1-\alpha)-\alpha\,, \tag{6.262}$$

and thus

$$\alpha=\frac{\tilde{p}^M}{1+\tilde{p}^M}\,. \tag{6.263}$$

In the special case $M=1$ that we will treat here explicitly, we obtain

$$g(y)=\frac{1}{1+\tilde{p}}\,,\quad \alpha=\frac{\tilde{p}}{1+\tilde{p}}\,. \tag{6.264}$$

We are left with evaluating

$$\frac{1}{\gamma}\int_{y_0}^{y_1}\frac{y}{1+y}\frac{1}{(y-\tilde{p})}dy=\frac{1}{\gamma}\left(\int_{y_0}^{y_1}\frac{(1-\alpha)dy}{1+y}+\int_{y_0}^{y_1}\frac{\alpha dy}{(y-\tilde{p})}\right)\,, \tag{6.265}$$

which immediately yields

$$(6.265)=\left(\frac{1}{\gamma}(1-\alpha)\ln(1+y)+\frac{1}{\gamma}\alpha\ln\mid y-\tilde{p}\mid\right)\Big|_{y_0}^{y_1}\,. \tag{6.266}$$

Inserting (6.256) and the integration limits,

$$t=0\leftrightarrow y_0=\tilde{A}+\tilde{p}\,, \tag{6.267}$$

$$t=\Delta\leftrightarrow y_1=\tilde{A}e^{\gamma\Delta}+\tilde{p}\,, \tag{6.268}$$

we find

$$
\begin{aligned}
(6.265) = & \frac{1}{\gamma}\frac{1}{1+\tilde{p}}\Big(\ln\left(1+\frac{A}{Q}\left(e^{\gamma\Delta}-1\right)^{-1}e^{\gamma\Delta}+p/Q\right) \\
& -\ln\left(1+\frac{A}{Q}\left(e^{\gamma\Delta}-1\right)^{-1}+p/Q\right)\Big) \\
& +\frac{1}{\gamma}\frac{\tilde{p}}{1+\tilde{p}}\left(\ln\left(\frac{A}{Q}\left(e^{\gamma\Delta}-1\right)^{-1}e^{\gamma\Delta}\right)-\ln\left(\frac{A}{Q}\left(e^{\gamma\Delta}-1\right)^{-1}\right)\right) \\
= & \, 2\pi/r. \qquad (6.269)
\end{aligned}
$$

After some elementary manipulations (using $\ln x+\ln y=\ln(xy)$), we obtain

$$
\begin{aligned}
& \frac{1}{\gamma}\frac{1}{1+p/Q}\ln\left(1+\frac{A/Q}{1+(A/Q)\left(e^{\gamma\Delta}-1\right)^{-1}+p/Q}\right)+\frac{p/Q}{1+p/Q}\Delta \\
& \quad = 2\pi/r\,. \qquad (6.270)
\end{aligned}
$$

This is the desired equation that allows us to determine the pulse interval Δ as a function of the various parameters. This implicit equation can only be solved numerically, but we are in a position to explicitly treat the limiting cases of large Q, corresponding to a linearization of S, and of small Q, where we expect saturation effects.

In the case $Q\to\infty$, we expand the l.h.s. of (6.270) into powers of $1/Q$, where we obtain in the leading approximation

$$
\frac{1}{\gamma}\frac{Ar}{Q}+\frac{pr}{Q}\Delta=2\pi\,, \qquad (6.271)
$$

which corresponds to our learlier result (5.47) by identifying

$$
Ar/Q \quad \text{with} \quad A \qquad (6.272)
$$

and

$$
pr/Q \quad \text{with} \quad c\,. \qquad (6.273)
$$

We now consider the case of small Q, or, equivalently, the case in which either the coupling between the neurons A or the external input becomes very large. To treat the first case,

$$
A\to\infty\,, \quad Q,p \quad \text{finite}\,, \qquad (6.274)
$$

we rewrite (6.270) in the form

$$
\frac{1}{\gamma}\frac{Q}{Q+p}\ln\left(1+\frac{1}{Q/A+\left(e^{\gamma\Delta}-1\right)^{-1}+p/A}\right)+\frac{p}{Q+p}\Delta=2\pi/r\,. \qquad (6.275)
$$

In the limit (6.274) it readily reduces to the simple result

$$
\Delta=2\pi/r\,. \qquad (6.276)
$$

The same result is obtained in the limiting case

$$p \to \infty \quad Q, A \quad \text{finite}: \quad \Delta = 2\pi/r\,. \tag{6.277}$$

The results (6.276) and (6.277) mean that for large inputs saturation of the neuron's response sets in, in which case the pulse intervals become independent of the size of the input. In the exercises, we will treat the case in which S (6.248) is used with $M = 2$, and indicate how to deal with the general case. So far, we determined Δ. With this parameter known, $\dot{\phi}$, and by mere integration over time, ϕ can be determined by means of (6.243).

Exercises. (1) Perform the above steps for S with $M = 2$.

Hint: The integrand in $\int \frac{x^2}{1+x^2}\frac{1}{x-d}dx$ can be decomposed into $\frac{x^2}{1+x^2}\frac{1}{x-d} = \frac{g(x)}{1+x^2} + \frac{\alpha}{x-d}$, where $\alpha = \frac{d^2}{1+d^2}$ and $g(x) = \frac{1}{1+d^2}(x+d)$. The integration yields $\frac{1}{1+d^2}\left(\frac{1}{2}\ln(1+x^2) + d\,\mathrm{arctg}\,x + d^2\ln(x-d)\right)$. Insert the lower and upper limits and discuss the limiting cases corresponding to $Q \to \infty$ and $A, p \to \infty$.

(2) Evaluate (6.259) for arbitrary, but integer $M \geq 3$.

Hint: Use (6.260) and $\frac{g(y)}{1+y^M} = \sum_{j=1}^{M}\frac{\alpha_j}{y-\beta_j}$, where $\beta_j = \exp(i3\pi j/(2M))$, $j = 1, 2, ..., M$. Why is $g(y)$ a polynomial of degree $M-1$?

(3) Use the results following (6.243) to determine $\dot{\phi}$ and ϕ explicitly for $M = 1$. Why is there a jump of $\dot{\phi}$ at $t = t_n$?

Hint: Observe (6.246).

7. Integrate and Fire Models (IFM)

The lighthouse model belongs to the large and important class of models that aim at studying phase locking and, in particular, synchronization among spiking neurons. In this and the following Chap. 8 we will deal with integrate and fire models (IFM) that serve the same purpose.

The basic differences between and commonalities of these models are as follows: The lighthouse model uses phases ϕ_k *and* dendritic currents as basic variables. The phases have only a formal meaning except that they determine the spike rate. They may grow to infinity. IFMs are based on model equations for the action potentials U_k of the neurons. These potentials are limited to a finite interval and, after firing, are reset to their origin. The general equations of Sect. 7.1 give a first clue to the relationship between these two kinds of models. Section 7.2 is devoted to a particularly simple example of an IFM. At least the first part of Sect. 7.3 (till (7.24)) is of general interest, because it shows how the relaxation of membrane potentials can be incorporated in lighthouse models.

7.1 The General Equations of IFM

The action potential U_k of neuron k obeys the equation

$$\frac{dU_k}{dt} = -\frac{U_k}{\tau} + I_k + I_k^{\text{ext}} \,. \tag{7.1}$$

The first term on the r.h.s. describes relaxation of U_k towards the resting potential with a relaxation time τ. In most models, the resting potential is set equal to zero and U_k is normalized so that it acquires its maximum value 1, that means

$$0 \leq U_k \leq 1 \,. \tag{7.2}$$

The next term in (7.1) is defined as follows:

$$I_k : \quad \textit{input from other neurons} \,. \tag{7.3}$$

It consists of the sum of the postsynaptic potentials (PSPs) that are triggered by pulses of afferent neurons at times t' (see (7.6)). I_k^{ext} represents external input stemming from sensory neurons

$$I_k^{\text{ext}} : \quad \textit{external input} \,. \tag{7.4}$$

A comparison between (7.1) and the equation of the phases as derived in (5.20) reveals that these equations have the same structure except for the first term on the r.h.s. in (7.1). Actually, the analogy between the models is quite close. A minor difference is this: While U_k adopts its values according to (7.2), ϕ_k runs in its first segment from 0 to 2π so that we arrive at the relationship

$$\phi_k = 2\pi U_k \,. \tag{7.5}$$

But in contrast to the integrate and fire models with (7.2), in the lighthouse model ϕ_k may increase indefinitely. We will discuss the relation between U_k and ϕ_k for $\phi_k > 2\pi$, i.e. in its higher segments, in (7.24).

Our considerations on the lighthouse model have taught us how to model the impact of the other neurons on the neuron under consideration by means of the processes within synapses and dendrites. When we eliminate the dynamics of the synapses and dendrites, we arrive at the relationship between the neuronal pulses and the input I_k. This relationship can be written in the form

$$I_k = \int_0^t K(t-t') \sum_{j \neq k} w_{kj} P_j(t') dt' \tag{7.6}$$

(see also (5.20)). The kernel K stems from the dynamics of the synapses and dendrites and will be specified below. w_{kj} represents synaptic strengths. The input in the form of pulses can be written as

$$P_j(t') = \sum_n \delta(t' - t_{j,n}) \,, \tag{7.7}$$

where δ is the well-known Dirac-function and $t_{j,n}$ represents the arrival times of the pulses of axon j. Because of the δ-functions, (7.6) reduces to

$$I_k = \sum_{j \neq k, n} w_{kj} K(t - t_{j,n}) \,, \tag{7.8}$$

where, depending on the dynamics of dendrites and synapses, K is given by

$$K(t) = \left\{ \begin{array}{ll} 0 & \text{for } t < 0 \\ e^{-\gamma' t} & \text{for } t \geq 0 \end{array} \right\} . \tag{7.9}$$

In this case, K is the Green's function of (5.1), see also (4.3). A general relationship for K that comes closer to physiological facts is given by

$$K(t) = \left\{ \begin{array}{ll} 0 & \text{for } t < 0 \\ c(\exp(-t/\tau_1) - \exp(-t/\tau_2)) & \text{for } t \geq 0 \end{array} \right\} , \tag{7.10}$$

where τ_1 and τ_2 are appropriately chosen relaxation times. We leave it as an exercise to the reader to determine a differential equation for which (7.10) is its Green's function. As will be further shown in the exercises, if we let $\tau_1 \to \tau_2$, a new kernel K can be obtained that is given by

$$K(t) = \begin{Bmatrix} 0 & \text{for } t < 0 \\ c\alpha^2 t \exp(-\alpha t) & t \geq 0 \end{Bmatrix} . \tag{7.11}$$

This K can be also obtained as the Green's function of (4.85) and is called Rall's α-function.

After having formulated the basic equations, we may discuss a method for their solution in the frame of IFM. The idea is to solve (7.1) with (7.8) explicitly until U_k reaches the value 1. This time fixes the initial time of a pulse from neuron k, whereupon the action potential of that neuron U_k is reset to zero. In practical applications, these equations have been solved numerically, which requires a considerable amount of computation time so that only few interacting neurons have been treated. In order to treat larger nets, considerable simplifications have been made. In the following, we will discuss a model due to Peskin that has been extended by Mirollo and Strogatz.

7.2 Peskin's Model

In the context of the synchronization of cardiac rhythms, Peskin derived a model that was later adopted to deal with synchronizing neurons by Mirollo and Strogatz. The activity of neuron k is described by a variable x_k, which obeys

$$\frac{dx_k}{dt} = S_0 - \gamma x_k \, , \quad k = 1, ..., N \tag{7.12}$$

and assumes the values

$$0 \leq x_k \leq 1 \, . \tag{7.13}$$

At an initial time, individual values x_k obeying (7.13) are prescribed. Then when time runs, because we assume $S_0 > 0$, according to (7.12) x_k will increase and reach the value 1, where a pulse is emitted by that neuron. Thereupon at an infinitesimally later time, t^+, the values x of all other neurons are increased by an amount ϵ or the other neuron fires when reaching $x_j = 1$. Thus, in short,

$$\begin{aligned} x_k(t) = 1 \to x_j(t^+) = x_j(t) + \epsilon \quad & \text{if} \quad x_j(t) + \epsilon < 1 \\ \text{or fires:} \quad & x_j = 1 \, . \end{aligned} \tag{7.14}$$

Peskin and later, in a more general frame, Mirollo and Strogatz could show that after some time all firings become synchronized. Let us try to make contact between this model and the lighthouse model on the one hand and the integrate and fire models on the other hand. Quite clearly, we have to identify

$$I_k^{\text{ext}} \leftrightarrow S_0 \tag{7.15}$$

and we will put

$$I_k^{\text{ext}} = I_0 \,, \tag{7.16}$$

i.e. all external signals are the same. Now we may use our knowledge about first-order differential equations that contain δ-functions (cf. Chap. 4). As we know, a jump of a variable can be caused by means of a δ-function. This leads us to make the hypothesis

$$I_k = \epsilon \sum_{j \neq k} \delta(t - t_j) \,. \tag{7.17}$$

For the sake of simplicity we further put

$$\gamma' = 1/\tau \,. \tag{7.18}$$

Clearly we will identify x_k with the action potential U_k. Thus the processes described by (7.12) and (7.14) are captured by

$$\frac{dU_k}{dt} = -\gamma' U_k + I_0 + \epsilon \sum_{j \neq k} \delta(t - t_j) \,. \tag{7.19}$$

According to Chap. 4 we know how the solution of (7.19) looks like, namely

$$U_k(t) = \int\limits_0^t e^{-\gamma'(t-\sigma)} \left(I_0 + \epsilon \sum_{j \neq k} \delta(\sigma - t_j) \right) d\sigma \,, \tag{7.20}$$

or, in a more explicit form,

$$U_k(t) = I_0 \frac{1}{\gamma'} \left(1 - e^{-\gamma' t} \right) + \epsilon \sum_{j \neq k} e^{-\gamma'(t - t_j)} H(t - t_j) \,, \tag{7.21}$$

where H is the Heaviside function introduced in (4.15). Since we will study the synchronized solution of (7.20) and its stability in great generality in Chap. 8, we will confine ourselves to one remark only, namely the determination of the phase-locked state. In this case, the neurons emit their pulses at the same time and according to the rule (7.14) their action potentials jump at the same time. This leads to the condition

$$U_k(t) = I_0 \frac{1}{\gamma'} \left(1 - e^{-\gamma' t} \right) + \epsilon N = 1 \,, \tag{7.22}$$

from which the time t, i.e. the time interval between pulses, can be deduced. Quite evidently, a sensitive result can be obtained only under the condition $\epsilon N < 1$. As we see by comparing (7.19) and (7.1) with (7.8), Peskin's model implies that the response via synapses and dendrites is without any delay, or, in other words, that the kernel K is a δ-function (see Exercise 3).

In the next section, we will study a model in which that response time is very long. This model has not yet been treated in the literature and those readers who would have fun studying a further model are referred to the next section.

7.3 A Model with Long Relaxation Times of Synaptic and Dendritic Responses

This section is devoted to those readers who are particularly interested in mathematical modelling. The first part of this section may also be considered as a preparation for Chap. 8. Our model bridges the gap between the lighthouse models that were treated in chaps. 5 and 6 and the integrate and fire models that we discussed at the beginning of the present chapter. To bring out the essentials, we again consider two interacting neurons. We describe their activities using the phase angles ϕ_1, ϕ_2. First of all, we generalize the lighthouse model by taking into account a damping of the rotation speed in between two firings. This means we want to mimic the effect of the damping term that occurs on the r.h.s. of (7.1) with (7.18). We are not allowed, however, to use the corresponding damping term $\gamma'\phi$ alone, because with increasing $\phi\,(> 2\pi)$ this expression would overestimate the effect of damping. Rather we have to take into account the fact that U is restricted to the region between 0 and 1, or correspondingly that the relaxation dynamics of ϕ in the intervals $[n2\pi, (n+1)2\pi]$ must be selfsimilar. In other words, we must reduce ϕ after each rotation. This is achieved by replacing ϕ with $\phi \bmod 2\pi$. In our present context, this mathematical expression means: if $\phi > 2\pi$, reduce it by an integer multiple of 2π so that $\phi' = \phi - 2\pi n$ with $0 \leq \phi' < 2\pi$. As can be seen by means of Fig. 5.11, this reduction can be achieved also with the help of a sum of Heaviside functions, H introduced in Sect. 4.1

$$\phi \bmod 2\pi = \phi - 2\pi \sum_{n\geq 1} H(\phi - 2\pi n)\,. \tag{7.23}$$

Thus when we "translate" the damping term on the r.h.s. of (7.1) into one for ϕ, we must apply the rules (7.5) *and* (7.23). This explains the meaning of the l.h.s. of the following

$$\dot{\phi}_1 + \gamma'\phi_1 - 2\pi\gamma' \sum_{n\geq 1} H(\phi_1 - 2\pi n)$$
$$= \sum_{n\geq 1} A(H(\phi_2 - 2\pi n) - H(\phi_1 - 2\pi(n+1)) + p\,. \tag{7.24}$$

The terms on the r.h.s. originate from the following mechanism: When the phase ϕ_2 of neuron 2 reaches $2\pi n$ that neuron fires and causes a change of the action potential of neuron 1. If the relaxation time γ of the synapses and dendrites is very large, we may assume that the corresponding term I_k in (7.1) does not decay. But then when ϕ_1 reaches 2π and the neuron is assumed to fire that influence on neuron 1 drops to zero. This is taken care of by a second term in the second sum in (7.24). p describes the external signal. A is a coupling coefficient. For neuron 2, we may write down an equation that is analogous to (7.24), namely

$$\dot{\phi}_2 + \gamma'\phi_2 = 2\pi\gamma' \sum_{n\geq 1} H(\phi_2 - 2\pi n) + \sum_{n\geq 1} A(H(\phi_1 - 2\pi n) - H(\phi_2 - 2\pi(n+1))) + p\,. \tag{7.25}$$

The phase-locked state obeys the

$$\dot{\phi} + \gamma'\phi = 2\pi\gamma' \sum_{n\geq 1} H(\phi - 2\pi n) + \sum_{n\geq 1} A(H(\phi - 2\pi n) - H(\phi - 2\pi(n+1))) + p\,. \tag{7.26}$$

In order to find its solution, we first rearrange the terms in the second sum according to

$$\sum_{n\geq 1} AH(\phi - 2\pi n) - \sum_{n\geq 2} AH(\phi - 2\pi n)\,, \tag{7.27}$$

which yields

$$(7.27) = A(H(\phi - 2\pi)) \tag{7.28}$$

and finally

$$(7.27) = A \quad \text{for} \quad \phi \geq 2\pi\,. \tag{7.29}$$

In this way, (7.26) can be transformed into

$$\dot{\phi} + \gamma'\phi = 2\pi\gamma' \sum_{n'\geq 1} H(\phi - 2\pi n') + p' \quad \text{with} \quad p' = p + A\,. \tag{7.30}$$

Let us briefly determine its solution with the initial condition

$$\phi(0) = \phi_0, \quad \text{where} \quad 0 \leq \phi_0 < 2\pi\,. \tag{7.31}$$

For a first interval, we obtain

$$n = 1: \quad 0 \leq t < t_1: \quad \phi(t) = p'/\gamma' + \alpha_0 e^{-\gamma t}\,. \tag{7.32}$$

The initial condition

$$t = 0: \quad p'/\gamma' + \alpha_0 = \phi_0 \tag{7.33}$$

fixes the constant

$$\alpha_0 = \phi_0 - p'/\gamma' \tag{7.34}$$

so that the solution reads

$$\phi(t) = p'/\gamma' + (\phi_0 - p'/\gamma')e^{-\gamma' t} . \tag{7.35}$$

We determine the first firing time by the requirement

$$t = t_1 \equiv \Delta_1 : \quad \phi(t_1) = 2\pi , \tag{7.36}$$

which yields

$$p'/\gamma' + (\phi_0 - p'/\gamma')e^{-\gamma t_1} = 2\pi , \tag{7.37}$$

or, explicitly,

$$e^{\gamma' \Delta_1} = \frac{p'/\gamma' - \phi_0}{p'/\gamma' - 2\pi} , \tag{7.38}$$

whereby we have to observe the requirement of a sufficiently strong external signal, namely

$$p'/\gamma' > 2\pi . \tag{7.39}$$

For the next interval, we obtain

$$n = 2 : \quad \Delta_1 \le t < t_2 : \quad 2\pi \le \phi < 4\pi , \quad H(\phi - 2\pi) = 1 . \tag{7.40}$$

By complete analogy with the just-performed steps, we obtain

$$\phi(t_2) = 2\pi + p'/\gamma' \left(1 - e^{-\gamma'(t_2 - t_1)}\right) , \tag{7.41}$$

and for the interval

$$t_2 - t_1 = \Delta_2 \tag{7.42}$$

the relation

$$\frac{1}{\gamma'}\left(1 - e^{-\gamma' \Delta_2}\right) = 2\pi/p' . \tag{7.43}$$

The time t_2 is determined by

$$t_2 = \Delta_2 + \Delta_1 . \tag{7.44}$$

For

$$\text{arbitrary} \quad n > 2 : \quad t_{n-1} \le t < t_n \tag{7.45}$$

we obtain

$$t_n = \Delta_1 + \Delta_2 + ... + \Delta_n = \Delta_1 + (n-1)\Delta , \tag{7.46}$$

where

$$\Delta_n = \Delta \quad \text{for} \quad n \geq 2 \,. \tag{7.47}$$

The general solution reads

$$\phi(t) = 2\pi(n-1) + p'/\gamma' \left(1 - e^{-\gamma'(t-t_{n-1})}\right) . \tag{7.48}$$

With the requirements

$$t = t_n, \quad \phi(t_n) = 2\pi n \,, \tag{7.49}$$

we obtain

$$\frac{1}{\gamma'}\left(1 - e^{-\gamma'\Delta_n}\right) = 2\pi/p' \,. \tag{7.50}$$

These simple considerations have allowed us to determine the explict form of the phase-locked state $\phi(t)$. In order to study its stability, we use the usual hypothesis

$$\phi_1 = \phi + \xi_1, \; \phi_2 = \phi + \xi_2 \,, \tag{7.51}$$

which again in the usual way leads, for example, to the following equation for $\xi_2(t)$

$$\begin{aligned} \dot{\xi}_2 + \gamma'\xi_2 = 2\pi\gamma' &\sum_{n\geq 1} \left(H(\phi + \xi_2 - 2\pi n) - H(\phi - 2\pi n)\right) \\ &+ A \sum_{n\geq 1} \left(H(\phi + \xi_1 - 2\pi n) - H(\phi - 2\pi n)\right) \\ &- A \sum_{n\geq 1} \left(H(\phi + \xi_2 - 2\pi(n+1)) - H(\phi - 2\pi(n+1))\right) \end{aligned} \tag{7.52}$$

that originates from the substraction of (7.26) from (7.25). Making the assumption that ξ_2 is a small quantity, we may expand the Heaviside functions occuring in (7.52) with respect to the arguments ξ_2 so that the conventional δ-functions appear. These δ-functions are formulated with arguments ϕ. When we proceed from them to those with arguments t_n, where the connection between t and ϕ is provided by (7.49), we obtain in the by now well-known way (cf. (5.112)–(5.115))

$$\begin{aligned} \dot{\xi}_2 + \gamma'\xi_2 = 2\pi\gamma' &\sum_{n\geq 1} \delta(t - t_n)\xi_2(t_n)/\dot{\phi}(t_n) \\ &+ A \sum_{n\geq 1} \delta(t - t_n)\xi_1(t_n)/\dot{\phi}(t_n) \\ &- A \sum_{n\geq 1} \delta(t - t_{n+1})\xi_2(t_{n+1})/\dot{\phi}(t_{n+1}) \,. \end{aligned} \tag{7.53}$$

An analogous equation can be derived for ξ_1 and we may then deduce equations for the sum $Z(t) = \xi_1 + \xi_2$ and the difference $\xi = \xi_2 - \xi_1$. The equations read

$$\begin{aligned}\dot{Z} + \gamma' Z &= \delta(t - t_1)(2\pi\gamma' + A)Z(t_1)/\dot{\phi}(t_1) \\ &\quad + \sum_{n\geq 2} \delta(t - t_n)2\pi\gamma' Z(t_n)/\dot{\phi}(t_1)\end{aligned} \tag{7.54}$$

and

$$\begin{aligned}\dot{\xi} + \gamma'\xi &= \delta(t - t_1)(2\pi\gamma' - A)\xi(t_1)/\dot{\phi}(t_1) \\ &\quad + \sum_{n\geq 2} \delta(t - t_n)(2\pi\gamma' - 2A)\xi(t_n)/\dot{\phi}(t_1) .\end{aligned} \tag{7.55}$$

The first term on the r.h.s. of (7.54) and (7.55), respectively, can be ignored if we let the process start at $t > t_1$. Then both equations bear a close resemblance to our previously derived (5.119), (5.121) with $B_1 = B_2 = 0$, where we may identify

$$a = 2\pi\gamma'/\dot{\phi} \quad \text{and} \quad -a = (2\pi\gamma' - 2A)/\dot{\phi} . \tag{7.56}$$

Quite clearly, these equations allow us to study the stability of the phase-locked state by complete analogy with Sect. 5.6. Then, for $t > t_1$, one may show that

$$Z(t_n + \epsilon) = Z(t_{n-1} + \epsilon) , \quad \epsilon \to 0 \tag{7.57}$$

up to higher orders of γ' in the frame of our approach, which neglects the jump of $\dot{\phi}(t_n)$ at $t_n \pm \epsilon$. If this jump is taken into account (see exercise), (7.57) is even exact. Stability of the phase-locked is secured if $A > 0$.

Our results shed new light on the question of stability, or, in other words, on the reasons why instability may occur. As we have seen above, there are two processes, namely the build-up of the action potential because of the terms with positive A in (7.24), (7.25) and then its decay caused by the negative term originating from the firing. In other words, if the build-up of the action potential cannot be compensated for strongly enough, an instability occurs (see also Exercise 6). Note that the damping constants γ, γ' in (5.121), (7.55) have quite different origins: γ stems from the damping of dendritic currents, while γ' from that of the action potential. In Chap. 8 we will study their combined effects.

We leave it as an exercise to the reader to generalize this section to N neurons.

Exercise 1. To which differential equation is

$$K(t) = \begin{cases} 0 & \text{for } t \leq 0, \\ a\left(e^{-\gamma_1 t} - e^{-\gamma_2 t}\right) & \text{for } t \geq 0 , \ \gamma_1 \neq \gamma_2 \end{cases}$$

the Green's function? Determine a!

Exercise 2. Derive (7.11) from (7.10) for $\tau_1 \to \tau_2$.
Hint: Put $c \propto (1/\tau_1 - 1/\tau_2)$.

Exercise 3. Show that (7.17) results from the elimination of dendritic currents in the lighthouse model for $\gamma \to \infty, \quad a_{jk} \to \infty, and \quad a_{jk}/\gamma$ finite.

Exercise 4. Prove (7.57) by means of Sect. 5.6.
Rederive (7.53), (7.55) taking into account the jump of $\phi(t_n \pm \epsilon)$, $\epsilon \to 0$.
Hint: Use (4.45)–(4.51) for $(t_0 - t_0^-) > 0, < 0$ separately. Use Sect. 4.3.

Exercise 5. Extend (7.24), (7.25) to many neurons.

Exercise 6. Convince yourself that without the second term (with $A > 0$) on the r.h.s. of (7.24) the driving force on ϕ_1 will increase indefinitely.

8. Many Neurons, General Case, Connection with Integrate and Fire Model

8.1 Introductory Remarks

The lighthouse model treated in Chaps. 5 and 6 had the advantage of simplicity that enabled us to study in particular the effects of delay and noise. In the present chapter we want to treat the more realistic model of Chap. 7 in detail. It connects the phase of the axonal pulses with the action potential U of the corresponding neuron and takes the damping of U into account. Furthermore, in accordance with other neuronal models, the response of the dendritic currents to the axonal pulses is determined by a second-order differential equation rather than by a first-order differential equation (as in the lighthouse model). The corresponding solution, i.e. the dendritic response, increases smoothly after the arrival of a pulse from another neuron. When the dendritic currents are eliminated from the coupled equations, we may make contact with the integrate and fire model. Our approach includes the effect of delays and noise. We will treat the first-order and second-order differential equations for the dendritic currents using the same formalism so that we can compare the commonalities of and differences between the results of the two approaches. The network connections of excitatory or inhibitory nature may be general, though the occurrence of the phase-locked state requires a somewhat more restricted assumption.

8.2 Basic Equations Including Delay and Noise

We denote the dendritic current of dendrite m by ψ_m and the axonal pulse of axon k by P_k. The equations for the dendritic currents read

$$\left(\frac{d}{dt}+\gamma\right)^{\alpha}\psi_m(t)=\sum_k a_{mk}P_k(t-\tau_{km})+F_{\psi,m}(t)\,, \tag{8.1}$$

where $\alpha = 1$ refers to the lighthouse model and $\alpha = 2$ to the integrate and fire model. Below we shall also discuss the extension of our approach to non-integer values of α. The constants and quantities in (8.1) have the following meaning: γ, damping constant; a_{mk}, coupling coefficient (synaptic strength)

τ_{km}, delay time; $F_{\psi,m}(t)$, fluctuating forces. The pulses P are expressed in the form

$$P(t) = f(\phi(t)), \tag{8.2}$$

where f is a periodic function of the phase ϕ that is strongly peaked and represented in the form

$$f(\phi(t)) = \dot{\phi} \sum_n \delta(\phi - 2\pi n), \tag{8.3}$$

or seen as a function of time t in the form

$$f(\phi(t)) = \tilde{f}(t) = \sum_n \delta(t - t_n). \tag{8.4}$$

The times t_n are defined by

$$t_n : \phi(t_n) = 2\pi n. \tag{8.5}$$

ϕ is interpreted as a phase angle that obeys

$$\dot{\phi}_j(t) + \gamma' \phi_j(t) \bmod 2\pi = S\left(\sum_m c_{jm} \psi_m \left(t - \tau'_{mj}\right) + p_{\text{ext,j}}, \Theta_j\right) + F_{\phi,j}(t). \tag{8.6}$$

Depending on

$$\gamma' = 0, \quad \text{or} \quad \gamma' = 1, \tag{8.7}$$

we are dealing with the lighthouse model in the first case and with the integrate and fire model in the second case, where we use the usual scaling of γ'. The connection between the phase angle ϕ and the action potential U is given by

$$\phi(t) \bmod 2\pi = 2\pi U(t). \tag{8.8}$$

We present the sigmoid function S that is well established by physiological experiments, e.g. Wilson [3], by the Naka–Rushton formula (5.7), or, in a rather good approximation, in the following form

$$S(X, \Theta) = 0 \quad \text{for} \quad X < X_{\min}, \tag{8.9}$$

$$= X \quad \text{for} \quad X_{\min} \le X \le X_{\max}, \tag{8.10}$$

$$= S_{\max} \quad \text{for} \quad X \ge X_{\max}. \tag{8.11}$$

The quantities in (8.6) have the following meaning: c_{jm}, coupling constants; τ'_{mj}, delay times; $p_{\text{ext},j}$, external signal; Θ_j, threshold; $F_{\phi,j}$, fluctuating forces.

8.3 Response of Dendritic Currents

In order to make contact with the more conventional representation of the integrate and fire model, we eliminate the dendritic currents. Equation (8.1) is of the form

$$\left(\frac{d}{dt}+\gamma\right)^{\alpha}\psi(t)=g(t)\,, \tag{8.12}$$

which may be supplemented by the initial conditions

$$\alpha=1:\ \psi(0)=0 \tag{8.13}$$

or

$$\alpha=2:\ \psi(0)=\dot{\psi}(0)=0\,. \tag{8.14}$$

Using a Green's function (see Sect. 4.4)

$$K_{\alpha}(t-\sigma)=(t-\sigma)^{\alpha-1}e^{-\gamma(t-\sigma)}\,, \tag{8.15}$$

the solution of (8.12) reads

$$\psi(t)=\int_{0}^{t}K_{\alpha}(t-\sigma)g(\sigma)d\sigma\,. \tag{8.16}$$

The Green's function possesses an important biophysical meaning: it represents the response of the dendritic current to a single spike – a δ-pulse – at time $t=\sigma$. According to the different values of α, the rise of the response may have various shapes (Figs. 4.11 and 4.12). In this way, the value of α can be experimentally determined and seems to be about 1.5, i.e. non-integer. Later we shall see how to deal mathematically with such non-integer values of α. In principle, (8.16) is a special solution of (8.12), whose general solution is found by adding to (8.16) a solution of the homogeneous equation (i.e. (8.12) with $g(t)\equiv 0$). In this way, any admitted initial condition on ψ can be realized. Because of (8.13) or (8.14) the solution of the homogenous equation vanishes so that (8.16) is already the general solution. Applying the relationship (8.16) to the general (8.1), and inserting the result into (8.6), yields the basic equations for the phases ϕ_j

$$\dot{\phi}_j(t)+\gamma'\phi_j(t)\bmod 2\pi=S\left(\sum_m c_{jm}\int_{0}^{t-\tau'_{mj}}K_{\alpha}\left(t-\tau'_{mj}-\sigma\right)\{...\}d\sigma\right.$$
$$\left.+p_{\mathrm{ext},j},\Theta_j\right)+F_{\phi,j}(t)\,, \tag{8.17}$$

where we abbreviated using the curly bracket

$$\{...\} = \sum_k a_{mk} P_k(\sigma - \tau_{km}) + F_{\psi,m}(\sigma) \,. \tag{8.18}$$

In the following we shall assume

$$a_{mk} \quad \text{time-independent}, \quad F = 0, \quad S \quad \text{linear} \tag{8.19}$$

and put

$$S\left(p_{\text{ext},j}, \Theta_j\right) = C_j \,. \tag{8.20}$$

Equation (8.17) can be rearranged to

$$\begin{aligned} &\dot{\phi}_j(t) + \gamma' \phi_j(t) \mathrm{mod} 2\pi \\ &\qquad = \sum_{mk} c_{jm} a_{mk} \times \int_{\tau'_{mj}}^{t} K_\alpha(t-\sigma) P_k(\sigma - \tau_{kmj}) d\sigma + C_j \end{aligned} \tag{8.21}$$

with

$$\tau_{kmj} = \tau_{km} + \tau'_{jm} \,. \tag{8.22}$$

In order not to overload our presentation, we make the following simplification

$$K_\alpha \to K \,. \tag{8.23}$$

Furthermore we introduce a relabeling

$$\tau_{kmj} \to \tau_\ell \,, \tag{8.24}$$

and

$$c_{jm} a_{mk} \to A_{jk,\ell} \,. \tag{8.25}$$

Note that these replacements do not mean any restriction. Equation (8.21) can be rewritten as

$$\dot{\phi}_j(t) + \gamma' \phi_j(t) \mathrm{mod} 2\pi = \sum_{k\ell} A_{jk,\ell} \int_0^t K(t-\sigma) f(\phi_k(\sigma - \tau_\ell)) d\sigma + C_j \,. \tag{8.26}$$

In the integral in (8.26) we replaced the lower limit by $\tau = 0$, which has no effect on our final results as may be shown by a more detailed analysis.

8.4 The Phase-Locked State

The phase-locked state is defined by

$$\phi_j = \phi, \quad j = 1, ..., N \tag{8.27}$$

and has to obey

$$\dot{\phi}(t) + \gamma' \phi(t) \text{mod} 2\pi = \sum_{k\ell} A_{jk,\ell} \int_0^t K(t - \sigma) f(\phi(\sigma - \tau_\ell)) d\sigma + C_j\,, \tag{8.28}$$

where we assume

$$C_j = C, \quad \sum_k A_{jk,\ell} = A_\ell \tag{8.29}$$

so that the r.h.s. of (8.28) becomes independent of j. As can be shown in a self-consistent way, under steady-state conditions, ϕ is periodic

$$\phi(t_n) = 2\pi n\,, \tag{8.30}$$

and we shall use the abbreviation

$$t_{n+1} - t_n = n\Delta, \quad n \text{ an integer.} \tag{8.31}$$

We establish the formal solution of (8.28) in the interval $t_n \leq t \leq t_{n+1}$ and put

$$\phi(t) = \phi(t_n) + \chi(t) \tag{8.32}$$

with

$$\chi(t_n) = 0 \text{ and } \dot{\chi} \geq 0\,. \tag{8.33}$$

Using (8.32), (8.31) and (8.33), we may transform (8.28) into

$$\dot{\chi}(t) + \gamma' \chi(t) = G(t)\,, \tag{8.34}$$

where G is an abbreviation of the r.h.s. of (8.28). Using the Green's function (8.15) with $\alpha = 1$ and replacing γ with γ', the solution of (8.34) reads

$$\chi(t) \equiv \phi(t) - \phi(t_n) = \sum_\ell A_\ell \int_{t_n}^t e^{-\gamma'(t-\sigma')} d\sigma' \int_0^{\sigma'} K(\sigma' - \sigma)$$

$$\times \sum_m \delta(\sigma - \tau_\ell - t_m) d\sigma + \int_{t_n}^t e^{-\gamma'(t-\sigma')} C d\sigma'\,. \tag{8.35}$$

In order to derive an equation for the interval Δ, we put $t = t_{n+1}$. Because of the δ-functions, we can easily evaluate the integrals (see Appendix 1). The solution reads

$$\phi(t_{n+1}) - \phi(t_n) = \sum_\ell A_\ell h_\alpha(\Delta, \gamma, \gamma', \tau'_\ell) + \frac{C}{\gamma'}\left(1 - e^{-\gamma'\Delta}\right), \tag{8.36}$$

where in the case of the lighthouse model $\alpha = 1$,

$$h_1(\Delta, \gamma, \gamma', \tau'_\ell) = \frac{e^{\gamma\tau'_\ell}}{\gamma' - \gamma} \cdot \frac{e^{-\gamma\Delta} - e^{-\gamma'\Delta}}{e^{\gamma\Delta} - 1} + \frac{1}{\gamma' - \gamma}\left(e^{-\gamma\Delta + \gamma\tau'_\ell} - e^{-\gamma'\Delta + \gamma'\tau'_\ell}\right), \tag{8.37}$$

$$\tau'_\ell = \tau_\ell \mathrm{mod}\Delta. \tag{8.38}$$

If

$$\gamma'\Delta \ll 1,\ \gamma\Delta \ll 1 \tag{8.38}$$

hold, (8.37) reduces to

$$h_1 = \frac{1}{\gamma}. \tag{8.39}$$

Because the Green's function K_2 can be obtained by differentiating K_1 with respect to γ, up to a factor -1, also h_2 can be obtained in this way,

$$h_2(\Delta, \gamma, \gamma', \tau_\ell) = -\frac{\partial}{\partial\gamma} h_1(\Delta, \gamma, \gamma', \tau_\ell). \tag{8.40}$$

In the case of (8.38) this reduces to

$$h_2 = \frac{1}{\gamma^2}. \tag{8.41}$$

Because due to (8.30) the l.h.s. of (8.36) is equal to 2π, (8.36) is an equation for Δ, which in the cases (8.38), (8.39) and (8.41) becomes particularly simple and coincides with our former results in Sect. 6.6. Also the case in which K_α (8.15) is defined for non-integers, α can be included using the results of Appendix 2 of Chap. 8. We thus obtain, again under the conditions of (8.38), $h_\alpha = (-1)^{1-\alpha}(d^{\alpha-1}/d\gamma^{\alpha-1})(1/\gamma)$, or, explicitly, $h_\alpha = \Gamma(\alpha - 1)\gamma^{-\alpha}$, where Γ is the usual Γ-function (which represents an extension of the factorial $n! = \Gamma(n)$ for an integer n to non-integer arguments (see Appendix 2)).

8.5 Stability of the Phase-Locked State: Eigenvalue Equations

In order not to overload our formulas, we first treat the case of a single delay time τ, where we put

$$\tau = M\Delta + \tau',\ 0 \le \tau' < \Delta,\ M \text{ an integer}, \tag{8.42}$$

and drop the index ℓ everywhere. For a linear stability analysis, we make the hypothesis

$$\phi_j(t) = \phi(t) + \xi_j(t) . \tag{8.43}$$

Inserting it into (8.26) and subtracting (8.28), we obtain

$$\begin{aligned} &\dot{\xi}_j(t) + \gamma'((\phi(t) + \xi_j(t))\mathrm{mod}2\pi - \phi(t)\mathrm{mod}2\pi) \\ &= \sum_k A_{jk} \int_0^t K(t,\sigma) \cdot \{f(\phi(\sigma') + \xi_k(\sigma')) - f(\phi(\sigma'))\} d\sigma , \end{aligned} \tag{8.44}$$

where $\sigma' = \sigma - \tau$. The evaluation of the curly bracket in (8.44), with $f(\phi)$ given by (8.3), in the limit of small $\mid \xi_k \mid$ can be done by close analogy with that of the r.h.s. of (5.109). Using the intermediate steps (5.110)–(5.115), we obtain (see also Exercise 1)

$$\begin{aligned} &\dot{\xi}_j(t) + \gamma' \left[(\phi(t) + \xi_j(t)) \bmod 2\pi - \phi(t) \bmod 2\pi\right] \\ &= \sum_k A_{jk} \int_0^t K(t,\sigma) \sum_n \frac{d}{d\sigma} \delta(\sigma - \tau - t_n) d\sigma \xi_k(t_n) \dot{\phi}(t_n)^{-1} . \end{aligned} \tag{8.45}$$

For a stationary phase-locked state, we may replace $\dot{\phi}(t_n)$ with $\dot{\phi}(t_0)$. It remains to evaluate the square bracket on the l.h.s. of (8.45) for small $\mid \xi_j \mid$. This was done in Sect. 7.3 with the result

$$[...] = \xi_j(t) - 2\pi \dot{\phi}(t_0)^{-1} \sum_{n=0} \delta(t - t_n) \xi_j(t_n) . \tag{8.46}$$

We are now in a position to write down an important first result of the transformed (8.44). Incidentally, to arrive at a concise presentation, we use on both sides of (8.44) the vector notation

$$\boldsymbol{\xi} = \begin{pmatrix} \xi_1 \\ \xi_2 \\ \vdots \\ \xi_L \end{pmatrix} , \tag{8.47}$$

and the matrix

$$\tilde{A} = \dot{\phi}(t_0)^{-1} (A_{jk}) . \tag{8.48}$$

On the r.h.s. of (8.45), we make use of (8.42) and put

$$t_{n'} = M\Delta + t_n , \tag{8.49}$$

or equivalently

$$n' = M + n . \tag{8.50}$$

Correspondingly, we replace n by $n'-M$ in (8.45). Finally, we drop the prime of n'. With the abbreviation

$$\hat{\gamma} = \gamma' 2\pi \dot{\phi}(t_0)^{-1} \tag{8.51}$$

we obtain, after a slight rearrangement, instead of (8.45) our final equation

$$\begin{aligned} \dot{\boldsymbol{\xi}} + \gamma' \boldsymbol{\xi} = \tilde{A} \int_0^t K(t,\sigma) \sum_n \frac{d}{d\sigma} \delta(\sigma - \tau' - t_n) \boldsymbol{\xi}(t_{n-M}) d\sigma \\ + \hat{\gamma} \sum_{n=0} \delta(t - t_n) \boldsymbol{\xi}(t_n) . \end{aligned} \tag{8.52}$$

We now turn to the solution of this equation and use, in a first step, the Green's function method to obtain

$$\begin{aligned} \boldsymbol{\xi}(t) = \tilde{A} \int_0^t e^{-\gamma'(t-s)} ds \int_0^s K(s,\sigma) \sum_n \frac{d}{d\sigma} \delta(\sigma - \tau' - t_n) \boldsymbol{\xi}(t_{n-M}) d\sigma \\ + \hat{\gamma} \int_0^t e^{-\gamma'(t-\sigma)} \sum_n \delta(\sigma - t_n) \boldsymbol{\xi}_n(t_n) d\sigma + \boldsymbol{\xi}_{hom}(t) , \end{aligned} \tag{8.53}$$

where $\boldsymbol{\xi}_{hom}(t)$ is a solution of the homogeneous (8.52). Clearly, to determine $\boldsymbol{\xi}(t)$ as a function of time, we need only the values of $\boldsymbol{\xi}$ at discrete times t_n. Therefore on both sides of (8.53) we put $t = t_n$. This converts (8.53) into a set of linear equations with time-independent coefficients. For the solution we can make the usual hypothesis

$$\boldsymbol{\xi}(t_n) = \boldsymbol{\xi}_0 \beta^n . \tag{8.54}$$

The evaluation of the first term (containing the double integral) on the r.h.s. of (8.53) depends on the explicit form of $K(t,\sigma)$. We first use (cf. (8.15) with $\alpha = 1$)

$$K(t,\sigma) \equiv K_1(t,\sigma) = e^{-\gamma(t-\sigma)} . \tag{8.55}$$

Because of the (derivative of the) δ-function, the integrals can easily be performed and we obtain for $t = t_N = N\Delta$, N an integer, $0 < \tau' < \Delta$,

$$\begin{aligned} R_1 \equiv \text{first term r.h.s. (8.53)} = -\tilde{A} \sum_{n=0}^{N-1} \boldsymbol{\xi}(t_{n-M}) \frac{1}{\gamma' - \gamma} \big(\gamma e^{-\gamma(t - t_n - \tau')} \\ - \gamma' e^{-\gamma'(t - t_n - \tau')} \big) . \end{aligned} \tag{8.56}$$

To evaluate the sum, we use the hypothesis (8.54), which yields

$$\begin{aligned} R_1 = -\tilde{A} \boldsymbol{\xi}_0 \beta^{-M} \bigg\{ \frac{\gamma}{\gamma' - \gamma} \cdot e^{\gamma\tau'} \frac{\beta^N - e^{-\gamma N \Delta}}{\beta e^{\gamma\Delta} - 1} \\ - \frac{\gamma'}{\gamma' - \gamma} e^{\gamma'\tau'} \frac{\beta^N - e^{-\gamma' N \Delta}}{\beta e^{\gamma'\Delta} - 1} \bigg\} . \end{aligned} \tag{8.57}$$

The evaluation of the second term in (8.53) is still simpler and yields

$$\hat{\gamma}\frac{\beta^N - e^{-\gamma' N\Delta}}{\beta e^{\gamma'\Delta} - 1}\boldsymbol{\xi}_0 \,. \tag{8.58}$$

Using the results (8.57) and (8.58), we obtain for (8.53) our final result

$$\begin{aligned}\boldsymbol{\xi}_0\beta^N = &- \tilde{A}\boldsymbol{\xi}_0\beta^{-M}\left\{\frac{\gamma}{\gamma' - \gamma}e^{\gamma\tau'}\frac{\beta^N}{\beta e^{\gamma\Delta} - 1}\right.\\ &\left. - \frac{\gamma'}{\gamma' - \gamma}e^{\gamma'\tau'}\frac{\beta^N}{\beta e^{\gamma'\Delta} - 1}\right\}\\ &+ \hat{\gamma}\frac{\beta^N}{\beta e^{\gamma'\Delta} - 1}\boldsymbol{\xi}_0 + \boldsymbol{\xi}_{hom}(t_N) + \text{terms independent of } \beta^N \,.\end{aligned} \tag{8.59}$$

This equation has been derived using $K = K_1$ in (8.53), as shown in (8.55). The result for $K = K_2(t, \sigma) = -dK_1/d\gamma$ can be immediately obtained by replacing R_1 with

$$R_2 = -dR_1/d\gamma \,. \tag{8.60}$$

We leave it as an exercise to the reader to perform the steps explicitly. In the case of non integer values of α in K_α, we can proceed correspondingly using fractal derivatives (see Appendix 2 to Chap. 8).

In order to see how to extract the eigenvalues β from (8.59), we consider a related example.

Exercise 1. Derive (8.45) in detail.

8.6 Example of the Solution of an Eigenvalue Equation of the Form of (8.59)

The determination of the eigenvalues β (or equivalently Γ with $\beta = e^{\Gamma}$) is somewhat tricky because we are dealing with an integral equation instead of the original differential equations for ψ and ϕ. To elucidate the problem, let us consider the differential equation

$$\dot{x} + \Gamma x = 0 \tag{8.61}$$

with its solution

$$x = x_0 e^{-\Gamma t} \,. \tag{8.62}$$

We write (8.61) in a different form

$$\dot{x} + \gamma x = -\gamma' x \tag{8.63}$$

with

$$\gamma + \gamma' = \Gamma . \tag{8.64}$$

Using the by now well-known Green's function, the formal solution of (8.63) reads

$$x(t) = \int_0^t e^{-\gamma(t-\sigma)}(-\gamma')x(\sigma)d\sigma + \alpha e^{-\gamma t} , \tag{8.65}$$

where $\alpha e^{-\gamma t}$ is the solution of the "homogeneous" (8.63). Because we are seeking a solution of the form of (8.62), we make the hypothesis

$$x(t) = \hat{x}_0 e^{-\hat{\Gamma} t} \tag{8.66}$$

and insert it into (8.65). The resulting integral can be easily solved and we obtain

$$\hat{x}_0 e^{-\hat{\Gamma} t} = \hat{x}_0(-\gamma')\frac{1}{\gamma - \hat{\Gamma}} e^{-\hat{\Gamma} t} + \hat{x}_0 \frac{\gamma'}{\gamma - \hat{\Gamma}} e^{-\gamma t} + \alpha e^{-\gamma t} . \tag{8.67}$$

As the reader will note, the structure of this equation is entirely analogous to that of (8.59), namely the r.h.sides contain terms β^N ($\leftrightarrow e^{\Gamma t}$) and terms of a different type. Thus when solving the "puzzle" of (8.67), we know how to deal with (8.59). Comparing the coefficients of $e^{\hat{\Gamma} t}$ or of $e^{-\gamma t}$ in (8.67), we obtain

$$1 = -\frac{\gamma'}{\gamma - \hat{\Gamma}} \tag{8.68}$$

and thus

$$\hat{\Gamma} = \gamma + \gamma' , \tag{8.69}$$

as well as

$$\alpha = -\hat{x}_0 \frac{\gamma'}{\gamma - \hat{\Gamma}} \tag{8.70}$$

and thus

$$\alpha = \hat{x}_0 . \tag{8.71}$$

In the context of (8.59), we are only interested in the eigenvalues β. Thus our example provides us with the rule: The terms $\boldsymbol{\xi}_{hom}(t)$ and *terms independent of* β^N cancel each other, and we need only to take terms with powers of β into account.

8.7 Stability of Phase-Locked State I: The Eigenvalues of the Lighthouse Model with $\gamma' \neq 0$

Dropping the last two terms in (8.59) and dividing both sides by β^N (N arbitrary), we obtain our fundamental eigenvalue equations (for the lighthouse model, $\alpha = 1$, but $\gamma' \neq 0$)

$$\begin{aligned} \boldsymbol{\xi}_0 = &- \tilde{A}\boldsymbol{\xi}_0\beta^{-M}\left\{\frac{\gamma}{\gamma' - \gamma}e^{\gamma\tau'}\frac{1}{\beta e^{\gamma\Delta} - 1}\right. \\ &\left. - \frac{\gamma'}{\gamma' - \gamma}e^{\gamma'\tau'}\frac{1}{\beta e^{\gamma'\Delta} - 1}\right\} + \hat{\gamma}\frac{1}{\beta e^{\gamma'\Delta} - 1}\boldsymbol{\xi}_0 \,. \end{aligned} \tag{8.72}$$

The most elegant way to solve this is to choose $\boldsymbol{\xi}_0$ as the solution to

$$\tilde{A}\boldsymbol{\xi}_\mu = \lambda_\mu\boldsymbol{\xi}_\mu \,, \tag{8.73}$$

where we distinguish the eigenvalues and eigenvectors by the index μ. Choosing $\boldsymbol{\xi}_0$ this way, we can transform (8.72) into

$$\begin{aligned} 1 = &- \lambda_\mu\beta^{-M}\left\{\frac{\gamma}{\gamma' - \gamma}e^{\gamma\tau'}\frac{1}{\beta e^{\gamma\Delta} - 1} - \frac{\gamma'}{\gamma' - \gamma}e^{\gamma'\tau'}\frac{1}{\beta e^{\gamma'\Delta} - 1}\right\} \\ &+ \hat{\gamma}\frac{1}{\beta e^{\gamma'\Delta} - 1} \,. \end{aligned} \tag{8.74}$$

Putting $\gamma' = \hat{\gamma} = 0$, we obtain the result for the original lighthouse model

$$1 = \lambda_\mu\beta^{-M}\frac{e^{\gamma\tau'}}{\beta e^{\gamma\Delta} - 1} \,, \tag{8.75}$$

from which we deduce the equation

$$\beta^{M+1} - \beta^M e^{-\gamma\Delta} = \lambda_\mu e^{\gamma\tau' - \gamma\Delta} \,. \tag{8.76}$$

We discussed the solutions of this type of equation in Sect. 5.7. Let us turn to (8.74). In order not to overload our discussion, we assume $\gamma' \neq \gamma$, $\tau' = 0$, and λ_μ is small. Because of the terms $\beta e^{\gamma\Delta} - 1$ and $\beta e^{\gamma'\Delta} - 1$ in the denominators, we seek three types of solutions that use the resulting singularities for

$$\beta_1 \text{ with } \beta_1 e^{\gamma\Delta} - 1 = \epsilon_1 \,, \tag{8.77}$$

$$\beta_2 \text{ with } \beta_2 e^{\gamma'\Delta} - 1 = \epsilon_2 \,, \tag{8.78}$$

$$\beta_3 \neq \beta_1, \beta_2 \,. \tag{8.79}$$

We assume ϵ_1 is small and obtain in (8.74) up to order ϵ_1

$$\epsilon_1 e^{-\gamma\Delta M} = -\lambda_\mu\frac{1}{\gamma' - \gamma}\left(\gamma - \frac{\gamma'\epsilon_1}{e^{(\gamma'-\gamma)\Delta} - 1}\right) + \hat{\gamma}\frac{\epsilon_1 e^{-\gamma\Delta M}}{e^{(\gamma'-\gamma)\Delta} - 1} \,, \tag{8.80}$$

or

$$\epsilon_1 = \left(e^{-\gamma \Delta M} - \frac{\lambda_\mu \cdot \gamma' + \hat{\gamma} e^{-\gamma \Delta M}}{e^{(\gamma' - \gamma)\Delta} - 1} \right)^{-1} \frac{\gamma \lambda_\mu}{\gamma - \gamma'} . \tag{8.81}$$

We may determine β_2 similarly and obtain

$$\epsilon_2 = \left(e^{-\gamma' \Delta M} + \lambda_\mu \frac{\gamma}{e^{(\gamma - \gamma')\Delta} - 1} \right)^{-1} \left(\lambda_\mu \frac{\gamma'}{\gamma' - \gamma} + \hat{\gamma} e^{-\gamma \Delta M} \right) . \tag{8.82}$$

In order to obtain the third class of roots, β_3, we assume that in a self-consistent manner

$$| \beta_3 | e^{\gamma' \Delta} \ll 1, \; | \beta_3 | e^{\gamma \Delta} \ll 1 . \tag{8.83}$$

Again for $\tau' = 0$, we obtain from (8.74)

$$\beta_3^M = \lambda_\mu (1 + \hat{\gamma})^{-1} , \tag{8.84}$$

which possesses the M roots

$$\beta_3 = \sqrt[M]{\lambda_\mu (1 + \hat{\gamma})^{-1} e^{2\pi i j / M}} , \quad j = 0, ..., M - 1 . \tag{8.85}$$

Because of these roots, oscillatory damping occurs. The above results can be further simplified, for instance for

$$e^{(\gamma' - \gamma)\Delta} - 1 \approx (\gamma' - \gamma)\Delta . \tag{8.86}$$

It is interesting to study the impact of the damping of the action potential, which is determined by γ', and to compare the eigenvalues β for $\gamma' \neq 0$ with those for $\gamma' = 0$. First of all, by $\gamma' \neq 0$ a new eigenvalue (connected with new eigenvectors) is introduced, namely β_2. At least for small $| \lambda_\mu |$, it leads to damping. The other eigenvalues β_1, β_3 correspond to (5.169), (5.170), respectively. In β_3, a $\gamma' \neq 0$ increases the damping, while the changes of β_2 due to $\gamma' \neq 0$ depend in a somewhat intricate way on λ_μ.

8.8 Stability of Phase-Locked State II: The Eigenvalues of the Integrate and Fire Model

We now turn to the integrate and fire model, i.e.

$$\alpha = 2 \quad \text{and} \quad \gamma' \neq 0 . \tag{8.87}$$

As mentioned above (cf. (8.60)), we can obtain the eigenvalue equation by differentiating the curly bracket on the r.h.s. of (8.74) with respect to γ, multiplied by (-1). Since the eigenvalues β are only weakly influenced by $e^{\gamma \tau'}$ and $e^{\gamma' \tau'}$, we put these factors equal to unity. After a slight rearrangement of terms, we obtain

$$1 = \beta^{-M+1}\lambda_\mu \left\{ \frac{\gamma'}{(\gamma'-\gamma)^2} \frac{\left(e^{\gamma'\Delta} - e^{\gamma\Delta}\right)}{\left(\beta e^{\gamma\Delta} - 1\right)\left(\beta e^{\gamma'\Delta} - 1\right)} - \frac{\gamma}{\gamma'-\gamma} \frac{\Delta}{\left(\beta e^{\gamma\Delta} - 1\right)^2} \right\} + \hat{\gamma} \frac{1}{\beta e^{\gamma'\Delta} - 1} . \tag{8.88}$$

In the cases $M = 0$ and $M = 1$, this is an equation of third order that can be solved using Cardano's formula. For $M > 1$, (8.88) must be solved numerically. In all cases, for $M \geq 0$, we can obtain the eigenvalues to a good approximation, however, provided γ' and γ are sufficiently far away from each other and $\hat{\gamma} \ll 1$, $\mid \lambda_\mu \gamma \Delta \mid$ and $\mid \lambda_\mu \gamma' \Delta \mid \ll \mid \gamma' - \gamma \mid$. Since the resulting expressions are still rather complicated, we first summarize the salient results. Again, as in the foregoing section, we obtain three classes of eigenvalues $\beta_1, \beta_2, \beta_3$, corresponding to (8.77), (8.78) and (8.85). The main qualitative difference consists in a splitting of β_1 into two eigenvalues, cf. (8.99). While β_1 (8.77) implied pure damping, the new eigenvalues β_1 (cf. (8.101)) imply oscillations provided $\lambda_\mu < 0$. Let us now derive and discuss the eigenvalues in more detail. In order not to overload our treatment of (8.88), we approximate

$$e^{\gamma'\Delta} - e^{\gamma\Delta} \quad \text{using} \quad (\gamma' - \gamma)\Delta . \tag{8.89}$$

Using this approximation in (8.88) and rearranging terms, we obtain

$$\left(1 - \hat{\gamma}\frac{1}{\beta e^{\gamma'\Delta} - 1}\right)\beta^{M-1} = \frac{\lambda\gamma'\Delta}{\gamma'-\gamma} \frac{1}{\left(\beta e^{\gamma\Delta} - 1\right)\left(\beta e^{\gamma'\Delta} - 1\right)} - \frac{\lambda\gamma\Delta}{\gamma'-\gamma} \frac{1}{\left(\beta e^{\gamma\Delta} - 1\right)^2} . \tag{8.90}$$

Incidentally, we drop the index μ of λ_μ and will add it only at the end. We seek the first kind of eigenvalues β_1 (for λ fixed) using the hypothesis

$$\beta_1 = e^{-\gamma\Delta}(1 + \epsilon_1) , \tag{8.91}$$

where ϵ_1 is a small quantity. Inserting (8.91) into (8.90) yields

$$\left(1 - \hat{\gamma}\frac{1}{e^{(\gamma'-\gamma)\Delta} - 1}\right) e^{-\gamma\Delta(M-1)}\epsilon_1^2 = \frac{\lambda\gamma'\Delta}{\gamma'-\gamma} \frac{\epsilon_1}{e^{(\gamma'-\gamma)\Delta}(1+\epsilon_1) - 1} - \frac{\lambda\gamma\Delta}{\gamma'-\gamma} . \tag{8.92}$$

Because of (8.51) and $\dot{\phi} \approx 2\pi/\Delta$, we put $\hat{\gamma} = \gamma'\Delta$. With (8.89), the bracket on the l.h.s. of (8.92) yields (up to higher order in $\gamma'\Delta, \gamma\Delta$)

$$-\frac{\gamma}{\gamma'-\gamma} . \tag{8.93}$$

Under the assumption

$$\epsilon_1 e^{(\gamma'-\gamma)\Delta} < \mid e^{(\gamma'-\gamma)\Delta} - 1 \mid \approx \mid \gamma' - \gamma \mid \Delta \tag{8.94}$$

we may expand the r.h.s. of (8.92) up to ϵ_1^2. After a rearrangement of terms and multiplying the resulting equation by $\gamma' - \gamma$, we finally obtain a quadratic algebraic equation of the form

$$a\epsilon_1^2 - b\epsilon_1 + c = 0 \,, \tag{8.95}$$

where

$$a = \lambda\gamma' \frac{e^{(\gamma'-\gamma)\Delta}}{(\gamma'-\gamma)^2\Delta} - \gamma e^{-\gamma\Delta(M-1)} \,, \tag{8.96}$$

$$b = \frac{\lambda\gamma'}{(\gamma'-\gamma)} \,, \tag{8.97}$$

$$c = \lambda\gamma\Delta \,. \tag{8.98}$$

Inserting (8.96)–(8.98) into the standard formula for the solution of (8.95), i.e.

$$\epsilon_1 = \frac{b}{2a} \pm \sqrt{\frac{b^2}{4a^2} - \frac{c}{a}} \tag{8.99}$$

yields rather lengthy expressions so that we prefer to discuss some general aspects. We note that there are two branches, $\pm$, of the solutions. Provided

$$\gamma' \ll \gamma \,, \tag{8.100}$$

(8.100) reduces in its lowest approximation to

$$\epsilon_1 = \pm\sqrt{\Delta\lambda}\, e^{\gamma\Delta(M-1)/2} \,. \tag{8.101}$$

Note that ϵ_1 becomes imaginary for $\lambda < 0$, and thus the eigenvalue β_1 becomes complex, leading to an oscillatory relaxation process. If, on the other hand,

$$\gamma \ll \gamma' \,, \tag{8.102}$$

the evaluation of (8.99) and its discussion become rather involved. The case $M = 1, \quad \gamma \to 0$, is, however, particularly simple and the quadratic equation (8.88) for β can be easily solved, yielding

$$\beta = 1 \pm \sqrt{\lambda\Delta}\, e^{-\gamma'\Delta/2} \,, \tag{8.103}$$

which, for $\lambda > 0$, implies an instability.

Let us turn to the second class of eigenvalues, where we put

$$\beta_2 = e^{-\gamma'\Delta}(1 + \epsilon_2) \,, \tag{8.104}$$

where ϵ_2 is assumed to be small. Because of the kind of singularity, it suffices to retain terms up to ϵ_2 in (8.88). The straightforward result reads

$$\epsilon_2 = \left(e^{-\gamma' \Delta(M-1)} + \frac{\lambda\gamma}{(\gamma' - \gamma)^3 \Delta} \right)^{-1} \times \left(\Delta e^{-\gamma' \Delta(M-1)} - \frac{\lambda}{(\gamma' - \gamma)^2} \right) \gamma' . \tag{8.105}$$

Again, a number of special cases may be discussed, which is left to the reader as exercise.

Finally, we discuss the third class of solutions β_3. Under the assumptions

$$\mid \beta_3 e^{\gamma' \Delta} \mid \ll 1 , \mid \beta_3 e^{\gamma \Delta} \mid \ll 1 \tag{8.106}$$

(8.90) reduces to

$$(1 + \hat{\gamma})\beta^{M-1} = \lambda \Delta \tag{8.107}$$

which possesses the by now well-known complex solutions

$$\beta = \sqrt[M-1]{(1 + \gamma' \Delta)^{-1} \lambda \Delta} \, e^{2\pi i j/(M-1)}, \, j = 0, ..., M - 2 \tag{8.108}$$

for $M > 1$. The cases $M = 0$ and $M = 1$ need not be considered here, because they are covered by the solutions β_1, β_2 (cubic equations!). In all the above results for $\beta_1, \beta_2, \beta_3$, we must finally replace λ by λ_μ. At any rate, in all cases, except (8.102), stability of the phase-locked state is guaranteed if $\mid \lambda_\mu \Delta \mid$ is small enough. This condition can be fulfilled if either the coupling between the neurons or $\Delta \propto \dot{\phi}^{-1}$ is small enough. $\dot{\phi}^{-1}$ small can be achieved if the external signal is large.

Exercise. Derive the eigenvalue equation corresponding to (8.88) for the kernel $K_\alpha(t, \sigma)$ with $\alpha = 1.5$.
Hint: Start from (8.74) and use the formalism of fractal derivatives (Appendix 2 to Chap. 8).

8.9 Generalization to Several Delay Times

In order to generalize the fundamental eigenvalue (8.88), that we derived for a single delay time to several delay times τ_ℓ, we have to start from (8.26). It is a simple matter to recognize that all we have to do are two things:

1) We put
$\tau_\ell = M_\ell \Delta + \tau'_\ell, M_\ell$ an integer, $0 \leq \tau'_\ell < \Delta$,
and neglect τ'_ℓ.

2) While in our derivation of (8.88) we considered only a single τ_ℓ in (8.26), we now take all terms of $\sum_\ell A_{jk,\ell} ...$ in (8.26) into account. We introduce the matrices

$$\tilde{A}_\ell = (A_{jk,\ell})\,\dot{\phi}(t_0)^{-1}\,. \tag{8.109}$$

The general eigenvalue equation reads

$$\boldsymbol{\xi}(0) = \Big[\sum_\ell \tilde{A}_\ell \Big\{\frac{\gamma'}{(\gamma'-\gamma)^2}\left(e^{\gamma'\Delta} - e^{\gamma\Delta}\right)\left(1-e^{\gamma'\Delta}\beta\right)^{-1}\left(1-e^{\gamma\Delta}\beta\right)^{-1} - \frac{\gamma\Delta}{\gamma'-\gamma}e^{\gamma\Delta}\left(1-e^{\gamma\Delta}\beta\right)^{-2}\Big\}\beta^{-M_\ell+1} - \hat{\gamma}\left(1-e^{\gamma'\Delta}\right)^{-1}\Big]\boldsymbol{\xi}(0)\,. \tag{8.110}$$

Since the vector $\boldsymbol{\xi}$ is high-dimensional, this matrix equation is high-dimensional also and cannot be solved explicitly except for special cases or by numerical procedures. A considerable simplification can be achieved, however, if all matrices can be simultaneously diagonalized. This is, for instance, the case if $\tilde{A}_\ell$ depends on the difference of the indices j and k

$$\tilde{A}_{jk,\ell} = \tilde{a}_\ell(j-k)\,. \tag{8.111}$$

In this case the eigenvectors of the corresponding matrix are plane waves under the assumption of periodic boundary conditions (cf. Sect. 6.9). We denote the eigenvalues of $\tilde{A}_\ell$ by $\lambda_{\mu\ell}$ and obtain from (8.110)

$$1 = \sum_\ell a_{\ell,1}\left(1-e^{\gamma'\Delta}\beta\right)^{-1}\left(1-e^{\gamma\Delta}\beta\right)^{-1}\beta^{-M_\ell+1} - \sum_\ell a_{\ell,2}\left(1-e^{\gamma\Delta}\beta\right)^{-2}\beta^{-M_\ell+1} - \hat{\gamma}\left(1-e^{\gamma'\Delta}\beta\right)^{-1} \tag{8.112}$$

with

$$a_{\ell,1} = \lambda_{\mu\ell}\frac{\gamma'}{(\gamma'-\gamma)^2}\left(e^{\gamma'\Delta}-e^{\gamma\Delta}\right), \tag{8.113}$$

$$a_{\ell,2} = \lambda_{\mu\ell}\frac{\gamma}{\gamma'-\gamma}\Delta e^{\gamma\Delta}\,. \tag{8.114}$$

The solution of (8.112) is equivalent to looking for the zeros of a polynomial.

8.10 Time-Dependent Sensory Inputs

In order to study the phase-locked state and its stability, we assumed that all sensory inputs, i.e.

$$C_j(t) \propto p_{\mathrm{ext}} \tag{8.115}$$

are equal and time-independent. Now we wish to discuss the case in which all inputs are still equal but time-dependent

$$C_j = C(t)\,. \tag{8.116}$$

Then the eigenvalues of the stability analysis allow us to estimate what happens if C varies in time. On general grounds we may state that if $C(t)$ changes slowly, compared to the time constants inherent in the eigenvalues, we may assume that phase locking persists, i.e. that the common phase follows adiabatically the changes of C. If C changes abruptly, complicated transients will occur that require the knowledge of the eigenvectors, which is, however, beyond the scope of the present book. If for some time all C_j have been equal and then suddenly adopt different values,

$$C_j = C \rightarrow \begin{cases} C_1 \\ C_2 \\ \vdots \\ C_N , \end{cases} \tag{8.117}$$

phase locking will break down. This is actually observed in experiments on phase locking between neurons of the visual cortex (cf. Chap. 3).

8.11 Impact of Noise and Delay

The combined influence of noise and delay was studied in detail in Chap. 6, where we examined the lighthouse model. The treatment of the integrate and fire model is basically the same, but the determination of the necessary eigenvectors is far more complicated and beyond the scope of this book. A few general comments may therefore suffice here.

By analogy with the result (6.202), we may expect that the time-dependence of the correlation function between neurons at times t_N and $t_{N'}$ is essentially determined by a superposition of powers of the eigenvalues, i.e. $\beta^{N-N'}$. While such delays may be experimentally observable, the determination of the coefficients may present a major problem and shall not be discussed here further.

8.12 Partial Phase Locking

In a complex neural network it may happen that only a subgroup of neurons becomes phase-locked, whereas other neurons are unlocked. So let us consider as an example a group of phase-locked neurons, whose indices we denote by J, K, and the remaining neurons, whose indices we denote by j', k'. In such a case, the original equations of the full network (without time delays)

$$\dot{\phi}_j(t) + \gamma' \phi_j(t) \mathrm{mod} 2\pi = \sum_k A_{jk} G\left(\phi_k(t)\right) + C_j + \hat{F}_j(t) \tag{8.118}$$

can be decomposed into the two groups of equations

$$\dot{\phi}_J(t) + \gamma' \phi_J(t) \bmod 2\pi = \sum_K A_{JK} G(\phi_K) + C_J + \hat{F}_J(t) + \sum_{k'} A_{Jk'} G(\phi_{k'}) \,, \tag{8.119}$$

where

$$\phi_J = \phi_K = \phi \,, \tag{8.120}$$

and

$$\dot{\phi}_{j'}(t) + \gamma' \phi_{j'}(t) \bmod 2\pi = \sum_{k'} A_{j'k'} G(\phi_{k'}) + C_{j'} + \hat{F}_{j'} + \sum_K A_{j'K} G(\phi) \,. \tag{8.121}$$

Since the neurons with indices j', k' are not phase-locked, in (8.119) they act as some kind of incoherent noise source on the phase-locked neurons. The noise source is represented by the last sum in (8.119). On the other hand, the last term in (8.121) acts as a coherent driving force. Depending on the coefficients $A_{j'K}$, this force may or may not be important for the behavior of the neurons with indices j'. If the latter force becomes too strong, one may expect that also the hitherto unlocked neurons will become locked to the first group and the last term in (8.119) will lose its character as a noise term. Clearly, our approach can include time-delays.

8.13 Derivation of Pulse-Averaged Equations

Our starting point is the set of (8.1)–(8.6), from which we will derive equations for quantities that are averages over a time interval T that contains several pulses. For the convenience of the reader we repeat the basic equations but drop the fluctuating forces $F_{\psi,m}, F_{\phi,j}$, because we assume that their time-averages vanish. The equations for the dendritic currents for $\alpha = 2$ (the integrate and fire model) then read

$$\left(\frac{d}{dt} + \gamma\right)^{\alpha} \psi_m(t) = \sum_k a_{mk} P_k(t - \tau_{km}) \,, \tag{8.122}$$

where the pulses are represented in the form

$$P_k(t) = f(\phi_k(t)) = \sum_n \delta(t - t_{nk}) \,, \tag{8.123}$$

where t_{nk} is the nth firing time of axon k. The equations for the phase angles ϕ_j read

$$\dot{\phi}_j(t) + \gamma' \phi_j(t) \bmod 2\pi$$
$$= S\left(\sum_m c_{jm}\psi_m(t - \tau'_{mj}) + p_{\mathrm{ext,j}}(t), \Theta_j\right) . \tag{8.124}$$

The average of the first term on the l.h.s. of (8.124) over time T yields

$$\frac{1}{T}\int\limits_t^{t+T} \dot{\phi}_j(t')dt' = \frac{1}{T}\left(\phi_j(t+T) - \phi_j(t)\right) , \tag{8.125}$$

or

$$N_j 2\pi/T \equiv 2\pi\omega_j , \tag{8.126}$$

where N_j is the number of axonal pulses in interval T and ω_j is the pulse rate. The size of the second term on the l.h.s. of (8.124) can be estimated as follows. Since, according to its definition,

$$\phi(t) \bmod 2\pi , \tag{8.127}$$

runs in the interval

$$(0, 2\pi) , \tag{8.128}$$

we obtain as an order of magnitude of the phase average

$$\overline{\phi(t) \bmod 2\pi} = \pi . \tag{8.129}$$

In the following we assume

$$\gamma'/2 \ll \omega_j , \tag{8.130}$$

which is fulfilled if the external input p is large enough, but S is not yet saturated.

We turn to the evaluation of the r.h.s. of (8.124). If the pulse rate changes only slowly, we may use the approximation

$$\overline{S(X)} \approx S(\overline{X}) , \tag{8.131}$$

where, in view of (8.124),

$$\overline{\psi_m}(t-\tau) = \frac{1}{T}\int\limits_{t-\tau}^{t-\tau+T} \psi_m(t')dt' = \frac{1}{T}\int\limits_t^{t+T} \psi_m(t'-\tau)dt' \tag{8.132}$$

and

$$\overline{p_{\mathrm{ext,j}}}(t) = \frac{1}{T}\int\limits_t^{t+T} p_{\mathrm{ext,j}}(t')dt' . \tag{8.133}$$

It remains to establish an equation for $\overline{\psi_m}$. To this end, using the Green's function $K_\alpha(t,\sigma)$, we transform (8.122) into

$$\psi_m(t) = \sum_k a_{mk} \int_0^t K_\alpha(t,\sigma) \sum_n \delta(\sigma - t_{nk}) d\sigma \,, \tag{8.134}$$

over which we take the average

$$\overline{\psi_m(t)} = \sum_k a_{mk} \frac{1}{T} \int_t^{t+T} dt' \int_0^{t'} K_\alpha(t',\sigma) \sum_n \delta(\sigma - t_{nk}) d\sigma \,. \tag{8.135}$$

(For simplicity we first treat the case $\tau_{km} = 0$.) Because of the δ-functions in (8.135), the integral over σ can be immediately evaluated so that we are left with considering

$$\frac{1}{T} \int_t^{t+T} dt' K_\alpha(t', t_{nk}) \,. \tag{8.136}$$

We use the fact that K_α is a peaked function. Choosing T large enough and neglecting boundary effects, we obtain

$$(8.136) = \begin{cases} \overline{K_\alpha} = \text{const. if} \quad t_{nk} \in (t, t+T) \\ \quad\;\; = 0 \qquad \text{otherwise}\,. \end{cases} \tag{8.137}$$

Making use of (8.135)–(8.137), we readily obtain

$$\begin{aligned} \overline{\psi_m(t)} &= \sum_k a_{mk} \overline{K}_\alpha \frac{1}{T} \, (\text{number of pulses of axon } k \text{ in } (t, t+T)) \\ &= \sum_k a_{mk} \overline{K}_\alpha \omega_k(t) \,. \end{aligned} \tag{8.138}$$

In order to correct for the neglect of τ_{km} in (8.134), we finally have to replace t with $t - \tau_{mk}$ on the r.h.s. of (8.138). This leads us to our final result

$$\overline{\psi_m(t)} = \sum_k a_{mk} \omega_k (t - \tau_{km}) \,. \tag{8.139}$$

In conclusion, we may formulate the equations for the pulse rates ω_j, where we use (8.125), (8.126), (8.130)–(8.133)

$$\omega_j(t) = \frac{1}{2\pi} S \left(\sum_m c_{jm} \overline{\psi_m}(t - \tau'_{mj}) + \overline{p}_{\text{ext},j}(t), \Theta_j \right) \,. \tag{8.140}$$

Equations (8.139) and (8.140) constitute our final result, which is a general form of the Jirsa–Haken equation that we will discuss in Chap. 10. It is an easy matter to eliminate the dendritic currents from (8.139), (8.140) by inserting (8.139) into (8.140)

$$\omega_j(t) = \frac{1}{2\pi} S\left(\sum_{mk} c_{jm} a_{mk} \omega_k(t - \tau_{km} - \tau'_{mj}) + \overline{p}_{\mathrm{ext},j}(t), \Theta_j\right) . \tag{8.141}$$

When properly interpreted, these equations lead to the Wilson–Cowan equations (cf. Chap. 10) under the assumption that the relaxation time of the axonal pulse rates is small.

Appendix 1 to Chap. 8: Evaluation of (8.35)

In the following we will be concerned with the evaluation of (8.35) and here with the first double integral for $t = t_{n+1}$. We first note an important property of the kernel

$$K_\alpha(t-\sigma) = (t-\sigma)^{\alpha-1} e^{-\gamma(t-\sigma)}\,, \tag{8.142}$$

which reads for $\alpha = 1$

$$K_1(t-\sigma) = e^{-\gamma(t-\sigma)}\,, \tag{8.143}$$

and for $\alpha = 2$

$$K_2(t-\sigma) = (t-\sigma)e^{-\gamma(t-\sigma)}\,. \tag{8.144}$$

Quite obviously, the kernel (8.144) can be obtained from (8.143) by means of a differentiation with respect to γ,

$$K_2 = -\frac{\partial K_1}{\partial \gamma}\,. \tag{8.145}$$

A similar relationship holds even if α is a non-integer number. In such a case, by use of a fractal derivative (see Appendix 2 to Chap. 8), we may write

$$K_\alpha = (-1)^{1-\alpha} \frac{d^{\alpha-1}}{d\gamma^{\alpha-1}} K_1, \quad \alpha \quad \text{an integer or non-integer}\,. \tag{8.146}$$

Because of (8.143) and (8.145), it will be sufficient to first consider only the following expression

$$\int_{t_n}^{t_{n+1}} e^{-\gamma'(t_{n+1}-\sigma')} d\sigma' \int_0^{\sigma'} e^{-\gamma(\sigma'-\sigma)} \underbrace{\sum_{n'} \delta(\sigma - \tau - t_{n'})}_{\sum_{n''} \delta(\sigma-\tau'-n''\Delta)} d\sigma\,. \tag{8.147}$$

In the second line in (8.147), we have chosen τ' and n'' in such a way that τ' lies within Δ. Exchanging the sequence of integrations and observing the integration regions,

$$t_n \le \sigma' \le t_{n+1} \tag{8.148}$$

and

$$0 \le \sigma \le \sigma' , \tag{8.149}$$

we may decompose (8.147) into

$$\int_0^{t_n} d\sigma \int_{t_n}^{t_{n+1}} d\sigma' + \int_{t_n}^{t_{n+1}} d\sigma \int_{\sigma}^{t_{n+1}} d\sigma' , \tag{8.150}$$

where the integrands are the same as in (8.147). We first evaluate the first double integral in (8.150), namely

$$(I) = \int_0^{t_n} d\sigma \int_{t_n}^{t_{n+1}} d\sigma' \sum_{n''} \delta(\sigma - \tau' - n''\Delta) e^{-\gamma'(t_{n+1}-\sigma')} e^{-\gamma(\sigma'-\sigma)} . \tag{8.151}$$

Performing the integration over σ' in (8.151) yields

$$(I) = \int_0^{t_n} d\sigma \sum_{n''} \delta(\sigma - \tau' - n''\Delta) \times e^{-\gamma' t_{n+1} + \gamma\sigma} \left[\frac{1}{\gamma' - \gamma} \left(e^{(\gamma'-\gamma)t_{n+1}} - e^{(\gamma'-\gamma)t_n} \right) \right] . \tag{8.152}$$

We introduce a new integration variable by

$$\sigma = \tilde{\sigma} + \tau' , \tag{8.153}$$

which transforms (8.152) into

$$(I) = \int_{-\tau'}^{t_n - \tau'} d\tilde{\sigma} \sum_{n''} \delta(\tilde{\sigma} - n''\Delta) e^{-\gamma' t_{n+1} + \gamma\tilde{\sigma} + \gamma\tau'} [...] , \tag{8.154}$$

where the square bracket is the same as in (8.152). Because of the δ-function, the integral in (8.154) can be immediately evaluated so that

$$(I) = \sum_{n''=0}^{(n-1)} e^{\gamma\Delta n''} e^{-\gamma' t_{n+1} + \gamma\tau'} [...] . \tag{8.155}$$

The sum in (8.155) is a geometric series that can immediately be evaluated to yield

$$(I) = \frac{e^{\gamma\Delta n} - 1}{e^{\gamma\Delta} - 1} e^{\gamma\tau'} \frac{1}{\gamma' - \gamma} \left(e^{-\gamma t_{n+1}} - e^{-\gamma t_n} e^{-\gamma'(t_{n+1}-t_n)} \right) , \tag{8.156}$$

from which, in the limit $t_n, t_{n+1} \to \infty$, only

$$(I) = \frac{e^{\gamma\tau'}}{e^{\gamma\Delta} - 1} \frac{1}{\gamma' - \gamma} \left(e^{-\gamma\Delta} - e^{-\gamma'\Delta} \right) \tag{8.157}$$

remains. We now evaluate the second part of (8.150), i.e.

$$(II) = \int_{t_n}^{t_{n+1}} d\sigma \int_{\sigma}^{t_{n+1}} d\sigma' \sum_{n''} \delta(\sigma - \tau' - n''\Delta) e^{-\gamma'(t_{n+1} - \sigma')} \cdot e^{-\gamma(\sigma' - \sigma)}, \tag{8.158}$$

which can be written as

$$(II) = e^{-\gamma' t_{n+1}} \int_{t_n}^{t_{n+1}} d\sigma \sum_{n''} \delta(\sigma - \tau' - n''\Delta) e^{\gamma\sigma} \int_{\sigma}^{t_{n+1}} d\sigma' \cdot e^{(\gamma' - \gamma)\sigma'}. \tag{8.159}$$

The evaluation of the second integral yields

$$(II) = e^{-\gamma' t_{n+1}} \int_{t_n}^{t_{n+1}} d\sigma \sum_{n''} \delta(\sigma - \tau' - n''\Delta) e^{\gamma\sigma} [...], \tag{8.160}$$

where the square bracket is defined by

$$[...] = \frac{1}{\gamma' - \gamma} \left(e^{(\gamma' - \gamma) t_{n+1}} - e^{(\gamma' - \gamma)\sigma} \right). \tag{8.161}$$

Observing the limits of integration

$$t_n \le \sigma = \tau' + n''\Lambda \le t_{n+1}, \tag{8.162}$$

we note that because of the δ-function the only contribution to this integral stems from

$$\sigma = t_n + \tau'. \tag{8.163}$$

Thus (8.160) is transformed into

$$(II) = e^{-\gamma' t_{n+1}} e^{\gamma(t_n + \tau')} \frac{1}{\gamma' - \gamma} \left(e^{(\gamma' - \gamma) t_{n+1}} - e^{(\gamma' - \gamma)(t_n + \tau')} \right). \tag{8.164}$$

In the limit

$$t_n, t_{n+1} \to \infty \tag{8.165}$$

this reduces to

$$(II) = \frac{1}{\gamma' - \gamma} \left(e^{-\gamma\Delta + \gamma\tau'} - e^{-\gamma'\Delta + \gamma'\tau'} \right). \tag{8.166}$$

With (8.157) and (8.166) our final result reads

$$(I) + (II) = \frac{e^{\gamma\tau'}}{e^{\gamma\Delta}-1}\frac{1}{\gamma'-\gamma}\left(e^{-\gamma\Delta} - e^{-\gamma'\Delta}\right) + \frac{1}{\gamma'-\gamma}\left(e^{-\gamma\Delta+\gamma\tau'} - e^{-\gamma'\Delta+\gamma'\tau'}\right) . \tag{8.167}$$

In the special case

$$\gamma\Delta \ll 1, \quad \gamma'\Delta \ll 1 \tag{8.168}$$

and thus

$$\gamma\tau' \ll 1, \quad \gamma'\tau' \ll 1 . \tag{8.169}$$

(8.167) readily reduces to

$$(8.167) = 1/\gamma . \tag{8.170}$$

Appendix 2 to Chap. 8: Fractal Derivatives

In the following we want to show how to introduce and deal with fractal derivatives at least as they are needed for the extension of the results of Chap. 8 to non-integer positive values of α in (8.17). In the following we will denote non-integers by α and β and we want to define

$$\frac{d^\alpha}{dx^\alpha} f(x) = \ ?\,. \tag{8.171}$$

We require that the result of the differentiation coincides with the usual results if α is an integer. Furthermore, we require

$$\frac{d^\alpha}{dx^\alpha}\frac{d^\beta}{dx^\beta} f(x) = \frac{d^{\alpha+\beta}}{dx^{\alpha+\beta}} f(x)\,. \tag{8.172}$$

A convenient function for our purposes is

$$f(x) = e^{-\gamma x}\,. \tag{8.173}$$

In accordance with the requirement (8.172), we define

$$\frac{d^\alpha}{dx^\alpha} e^{\gamma x} = (-\gamma)^\alpha e^{-\gamma x}\,. \tag{8.174}$$

By means of (8.173), we can construct other functions, e.g.

$$\frac{1}{x} = \int\limits_0^\infty e^{-\gamma x} d\gamma\,, \quad x > 0\,. \tag{8.175}$$

The application of the rule (8.174) transforms (8.175) into

$$\frac{d^\alpha}{dx^\alpha}\frac{1}{x} = \int\limits_0^\infty (-\gamma)^\alpha e^{-\gamma x} d\gamma\,. \tag{8.176}$$

Introducing the new variable

$$y = -\gamma x\,, \tag{8.177}$$

(8.176) is transformed into

$$\frac{d^\alpha}{dx^\alpha}\frac{1}{x} = x^{-\alpha-1}\int\limits_{-\infty}^{0} y^\alpha e^y dy\,. \tag{8.178}$$

By partial integration of the integral in (8.178), we obtain for $\alpha > 1$

$$\int\limits_{-\infty}^{0} y^\alpha e^y dy = y^\alpha e^y\,|_{-\infty}^{0} - \alpha\int\limits_{-\infty}^{0} y^{\alpha-1} e^y dy\,. \tag{8.179}$$

Introducing the abbreviation

$$\int\limits_{-\infty}^{0} y^\alpha e^y dy = \tilde{\Gamma}(\alpha)\,, \tag{8.180}$$

we thus obtain the recursion relation

$$\tilde{\Gamma}(\alpha) = -\alpha\tilde{\Gamma}(\alpha-1)\,. \tag{8.181}$$

This is strongly reminiscent of the recursion relation for the factorial

$$n! = \int\limits_{0}^{\infty} x^n e^{-x} dx\,, \tag{8.182}$$

where by partial integration we find

$$n! = -x^n e^{-x}\,|_0^\infty + n\int\limits_{0}^{\infty} x^{n-1} e^{-x} dx\,, \tag{8.183}$$

or, denoting the r.h.s. of (8.182) by $\Gamma(n)$ the relation

$$\Gamma(n) = n\Gamma(n-1)\,, \tag{8.184}$$

or for noninteger α

$$\Gamma(\alpha) = \alpha\Gamma(\alpha-1)\,. \tag{8.185}$$

Γ is nothing but the conventional Γ-function. In order to connect (8.181) and (8.185), we put

$$\tilde{\Gamma}(\alpha) = (-1)^\alpha\Gamma(\alpha)\,. \tag{8.186}$$

Using (8.186) in (8.178), we find our final formula for the derivatives of 1/x

$$\frac{d^\alpha}{dx^\alpha}\frac{1}{x} = (-1)^\alpha\Gamma(\alpha)x^{-\alpha-1}\,. \tag{8.187}$$

Note that within our formalism (8.187) may become complex because of $(-1)^\alpha$. This is, however, a fully consistent formalism for our present purposes. In this way also derivatives of any negative power of x can be obtained.

Another fractal derivative needed in Chap. 8 is that of the function

$$\frac{1}{x}e^{\lambda x} = \int_{-\infty}^{\lambda} e^{\gamma x} d\gamma, \quad x > 0\,, \tag{8.188}$$

where the l.h.s. has been expressed by the integral on the r.h.s. Proceeding as before, we obtain

$$\frac{d^\alpha}{dx^\alpha}\left(\frac{1}{x}e^{\lambda x}\right) = \int_{-\infty}^{\lambda} \gamma^\alpha e^{\gamma x} d\gamma\,, \tag{8.189}$$

and using a new integration variable

$$\frac{d^\alpha}{dx^\alpha}\left(\frac{1}{x}e^{\lambda x}\right) = x^{-\alpha-1}\int_{-\infty}^{\lambda/x} y^\alpha e^y dy\,. \tag{8.190}$$

The integral of the r.h.s. now obeys the recursion relation

$$Z(\alpha,\Lambda) = \Lambda^\alpha e^\Lambda - \alpha Z(\alpha-1,\Lambda) \tag{8.191}$$

as can be checked by partial integration. We have put

$$\Lambda = \lambda/x\,. \tag{8.192}$$

By use of (8.191), any function Z for arbitrary α can be reduced to an expression that only contains

$$Z(\beta,\Lambda), \quad 0 \le \beta < 1\,. \tag{8.193}$$

The recursion relation (8.191) can further be simplified by the hypothesis

$$Z(\alpha,\Lambda) = e^\Lambda G(\alpha,\Lambda) \tag{8.194}$$

that transforms (8.191) into

$$G(\alpha,\Lambda) = \Lambda^\alpha - \alpha G(\alpha-1,\Lambda)\,. \tag{8.195}$$

Exercise. Determine the fractal derivatives of

$(x-c)^{-n}, \quad n > 0$ and an integer,
$(x-c)^{-1}e^{-ax}$,
$(x-c_1)^{-1}(x-c_2)^{-1}e^{-ax}$.

Hints: Write $(d^\alpha/dx^\alpha)(1/x^n)$ as $(d^\alpha/dx^\alpha)(d^n/dx^n)(1/x)(-1)^n$.
Decompose the product of the fractions into a sum.

Part III

Phase Locking, Coordination and Spatio-Temporal Patterns

9. Phase Locking via Sinusoidal Couplings

This chapter deals with biological systems composed of elements whose states can be described by phase variables. It will be assumed that the coupling between the elements is determined by sine-functions of their relative phases. As examples of such systems, we treat two neurons (Sect. 9.1), chains of neurons in the lamprey (Sect. 9.2), correlated movements between index fingers (Sect. 9.3) as a paradigm of limb coordination, and, more generally, quadruped motion (Sect. 9.4). In Sect. 9.5 we return to the neuronal level dealing with neuronal groups. As we will see, all these cases share a common mathematical ground.

9.1 Coupling Between Two Neurons

As we outlined in Chap. 2, many types of neurons generate an ongoing spike train, or a sequence of spike bursts in response to a constant stimulus. In Chap. 5, we studied the interaction between two such neurons, whereby each neuron was modelled as a phase oscillator. The formulation for the interaction between the two neurons took into account the actual spike trains as well as the dynamics of the dendritic currents. There is a still simpler approach to the treatment of the coupling between neurons that are again described by phase oscillators, but where the coupling is a smooth periodic function of the phase differences between the two neurons. The modelling of coupled nonlinear oscillators by means of phase oscillators with sinusoidal phase-coupling has a long history. It appeared in radio engineering, later in laser physics and was applied to chemical waves by Kuramoto and to the modelling of neurons by Cohen and others. As has been shown in the literature, such a coupling can be quite a good approximation for neuronal interaction under specific circumstances. Here we will consider such a phase-coupling model at two levels, namely at a level that refers to spiking neurons, and at a level where we are dealing with spike *rates*. These latter equations are special cases of the spike averaged equations we derived in Sect. 8.13. Let us elucidate the corresponding formalism that applies to both cases in more detail. The phases of the phase oscillators obey the equations

$$\frac{d\phi_j}{dt} = \omega_j \,, \quad j = 1, 2 \,, \tag{9.1}$$

where ω_j is the rotation speed of the phase. The coupling between the two oscillators is taken care of by coupling functions W that depend on the phase difference. Thus (9.1) has to be replaced by

$$\frac{d\phi_1}{dt} = \omega_1 + W_1(\phi_2 - \phi_1) \tag{9.2}$$

and

$$\frac{d\phi_2}{dt} = \omega_2 + W_2(\phi_1 - \phi_2) . \tag{9.3}$$

Since the processes under consideration are periodic in their phases, we assume

$$W_j(\phi + 2\pi) = W_j(\phi) . \tag{9.4}$$

We introduce the phase difference

$$\phi = \phi_2 - \phi_1 \tag{9.5}$$

and subtract (9.2) from (9.3),

$$\frac{d\phi}{dt} = \omega_2 - \omega_1 + W_2(\phi) - W_1(-\phi) . \tag{9.6}$$

Any periodic function can be represented as a Fourier series and in the present context it has turned out that it is sufficient to use the first term only. Thus we approximate W by means of

$$W_j = a_j \sin(\phi + \sigma_j) . \tag{9.7}$$

Using (9.7), the last two terms of (9.6) can be written as

$$a_2 \sin(\phi + \sigma_2) - a_1 \sin(-\phi + \sigma_1) . \tag{9.8}$$

By means of the mathematical identity

$$\sin(\alpha + \beta) = \sin\alpha \cos\beta + \cos\alpha \sin\beta , \tag{9.9}$$

and a rearrangement of terms, this expression can be cast into the form

$$(9.8) = (a_2 \cos\sigma_2 + a_1 \cos\sigma_1) \sin\phi + (a_2 \sin\sigma_2 - a_1 \sin\sigma_1) \cos\phi . \tag{9.10}$$

Now we wish to make use of (9.9) in the reverse direction, namely we wish to write

$$(9.10) = A\cos\phi + B\sin\phi \overset{!}{=} \alpha \sin(\phi + \delta) , \tag{9.11}$$

whereby the coefficients A and B can be immediately found by comparing the prefactors of $\sin\phi$ and $\cos\phi$. In order to apply (9.9), we put

$$A = \alpha \sin\delta , \quad B = \alpha \cos\delta , \tag{9.12}$$

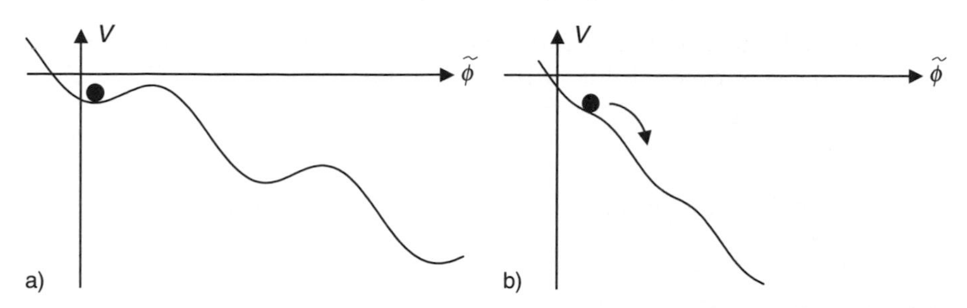

Fig. 9.1. The potential function $V(\tilde{\phi}) = -\Delta\tilde{\phi} - \alpha\cos\tilde{\phi}$: **a)** for $|\alpha| > \Delta$, example $\Delta, \alpha > 0$; **b)** for $|\alpha| < \Delta$, example $\Delta, \alpha > 0$

from which we deduce

$$\alpha^2 = A^2 + B^2, \quad \text{or} \quad \alpha = \sqrt{A^2 + B^2}\,, \tag{9.13}$$

as well as

$$A/B = \operatorname{tg}\delta, \quad \text{or} \quad \delta = \arctan(A/B)\,. \tag{9.14}$$

Making use of (9.9), (9.12), as well as of (9.11), we can cast (9.6) with (9.7) into the final form

$$\frac{d\phi}{dt} = \Delta + \alpha\sin(\phi + \delta)\,, \tag{9.15}$$

where

$$\Delta = \omega_2 - \omega_1\,. \tag{9.16}$$

Our final step consists of introducing a new variable according to

$$\tilde{\phi} = \phi + \delta\,. \tag{9.17}$$

In this way, we cast (9.15) into the final form

$$\frac{d\tilde{\phi}}{dt} = \Delta - \alpha\sin\tilde{\phi} = -\frac{\partial V}{\partial\tilde{\phi}}\,. \tag{9.18}$$

In the last equality, we have introduced a potential function V that allows us to discuss the behavior of the phase angle $\tilde{\phi}$ in a simple manner (see Fig. 9.1). There are two ranges of behavior of $\tilde{\phi}$.

If

$$|\,\alpha\,| > \Delta: \quad \text{phase locking} \tag{9.19}$$

holds, we find a time-independent solution to (9.18), because the term Δ can be compensated by the last term $\alpha\sin\tilde{\phi}$. If, however,

$$|\,\alpha\,| < \Delta: \quad \text{free running} \tag{9.20}$$

holds, we find a free-running solution in which the phase difference increases over the course of time (cf. Fig. 9.1b). In the first case we immediately find

$$\tilde{\phi} = \arc \sin \frac{\Delta}{\alpha} . \tag{9.21}$$

As is evident from Fig. 9.1a, the phase-locking solution is stable.

9.2 A Chain of Coupled-Phase Oscillators

In this section, we present an example that refers to the spike rate level and deals with the swimming motion of a lamprey, an eel-like vertebrate. This problem has been examined by several authors (see references) and we follow the nice presentation by Wilson (1999). The lamprey swims by generating travelling waves of neural activity that pass down its spinal cord. The modelling rests on the concept of what is called "a central pattern generator". It is defined as a neural network that produces stereotypical limit cycle oscillations in response to a constant spike rate input. The latter is also called "the command signal". The signal triggers the oscillations and determines their frequency. Its experimental counterpart can be found in the work of Shik et al. They showed that decerebrate cats could walk or trot when their spinal cord was excited by a constant electric input signal. Experimentally in the case of the lamprey, it has been shown that an intact spinal cord isolated from the rest of the body will generate rhythmic bursts of neural activity appropriate for swimming in response to constant stimulation.

Further experiments have shown that even small sections of the spinal cord are capable of generating rhythmic bursts of spikes in which activity on one side alternates with that on the other. Such oscillatory networks in the spinal segments cause alternate contraction and relaxation of the body muscles on opposite sides of the body during swimming. Here we will not go into the details of the local neural network, rather we wish to model the occurrence of travelling waves by means of coupled-phase oscillators. We consider only nearest-neighbor couplings. We then define a travelling wave as a phase-locked solution of these equations with a constant phase difference between the adjacent segments. Using a synaptic delay σ and denoting the phases by Θ, the basic equations are given by

$$\frac{d\Theta_1}{dt} = \omega_1 + a_a \sin(\Theta_2 - \Theta_1 + \sigma) , \tag{9.22}$$

$$\begin{aligned} \frac{d\Theta_i}{dt} &= \omega_i + a_a \sin(\Theta_{i+1} - \Theta_i + \sigma) \\ &\quad + a_d \sin(\Theta_{i-1} - \Theta_i + \sigma) , \quad i = 2, ..., N-1 , \end{aligned} \tag{9.23}$$

$$\frac{d\Theta_N}{dt} = \omega_N + a_d \sin(\Theta_{N-1} - \Theta_N + \sigma) , \tag{9.24}$$

where a_a represents the strength of the coupling in an ascending sequence of neurons (tail → head), whereas a_d represents the corresponding couplings for the descending sequence. A pairwise subtraction of these equations and using

$$\phi_i = \Theta_{i+1} - \Theta_i \tag{9.25}$$

leads to the equations

$$\frac{d\phi_1}{dt} = \Delta\omega_1 + a_a[\sin(\phi_2 + \sigma) - \sin(\phi_1 + \sigma)] + a_d \sin(-\phi_1 + \sigma), \tag{9.26}$$

$$\begin{aligned} \frac{d\phi_i}{dt} &= \Delta\omega_i + a_a[\sin(\phi_{i+1} + \sigma) - \sin(\phi_i + \sigma)] \\ &\quad + a_d[\sin(-\phi_i + \sigma) - \sin(-\phi_{i-1} + \sigma)], \end{aligned} \tag{9.27}$$

$$\begin{aligned} \frac{d\phi_{N-1}}{dt} &= \Delta\omega_{N-1} - a_a \sin(\phi_{N-1} + \sigma) \\ &\quad - a_d[\sin(-\phi_{N-1} + \sigma) - \sin(-\phi_{N-2} + \sigma)] \end{aligned} \tag{9.28}$$

with the abbreviation

$$\Delta\omega_i = \omega_{i+1} - \omega_i . \tag{9.29}$$

It is assumed that $\Delta\omega_i$ is determined by command signals coming from the brain. In principle, many different states are possible, but here we are interested in the travelling wave solution that is defined by

$$\frac{d\phi_i}{dt} = 0,\ \phi_i = \phi \quad \text{for all} \quad i . \tag{9.30}$$

By this assumption, the equations (9.26)–(9.29) are converted into

$$\Delta\omega_1 + a_d \sin(-\phi + \sigma) = 0 , \tag{9.31}$$

$$\Delta\omega_i = 0, \quad i = 2, ..., N-2 , \tag{9.32}$$

$$\Delta\omega_{N-1} - a_a \sin(\phi + \sigma) = 0 . \tag{9.33}$$

We note that $\Delta\omega_1$ and $\Delta\omega_{N-1}$ cannot both be zero unless $\sigma = 0$ or π. Therefore, as an example, we choose

$$\Delta\omega_{N-1} = 0 \tag{9.34}$$

and are then left with

$$\Delta\omega_1 = -a_d \sin(-\phi + \sigma) \tag{9.35}$$

as well as

$$\phi = -\sigma . \tag{9.36}$$

From (9.35) we conclude

$$\Delta\omega_1 = \omega_2 - \omega_1 = -a_d \sin\sigma\,, \tag{9.37}$$

i.e. ω_1 must be slightly larger than all other ω_i. In this way, we obtain forward swimming. Backward swimming can be achieved using $\omega_1 = 0$. One may show that stability is secured if

$$a_a > 0, \quad a_d > 0, \quad 0 < \pi/4 \tag{9.38}$$

holds. Thus a constant command signal to all except the first or last segment generates a travelling wave, and by sending a slightly larger signal to the first or last the direction of swimming can be determined.

In this analysis we have been interested in deriving travelling waves. There is yet an additional requirement, namely that the travelling wave of neural activity in the spinal cord has a wavelength equal to the body length of the animal. This can be achieved by choosing $\Delta\omega_1$ and $\Delta\omega_{N-1}$ not equal to zero. Using the fact that the number of segment equations N is equal to the number of segments, $N = 100$, the phase lag between segments must be $\phi = \pi/50$, which yields, using (9.31) and (9.33), the relations

$$\Delta\omega_1 = -a_d \sin(-\pi/50 + \sigma) \tag{9.39}$$

and

$$\Delta\omega_{N-1} = +a_a \sin(\pi/50 + \sigma)\,. \tag{9.40}$$

Since the coupling strength a_a and the phase shift σ are given, the frequencies can now be determined.

9.3 Coupled Finger Movements

When humans walk or run, or horses walk, trot or gallop, these movements are characterized by well coordinated movements of the individual limbs so that we may speak of movement *patterns*. In one way or another, these patterns reflect activity patterns of corresponding neurons in the brain, in particular the cerebellum, and in the spinal cord. A particularly fascinating aspect is the abrupt change between gaits. Haken (1983) suggested at an early stage that these changes closely resemble phase transitions of synergetic systems. In the study of such phase transitions, experiments by Kelso et al. on changes in coordinated finger movements have played a pivotal role. In the following we present the corresponding Haken–Kelso–Bunz model. In his experiments, Kelso asked a subject to move his (or her) index fingers in parallel at a comparatively slow speed (frequency). When the prescribed frequency was increased more and more, suddenly and involuntarily, the subject switched to a symmetric movement (Fig. 9.2).

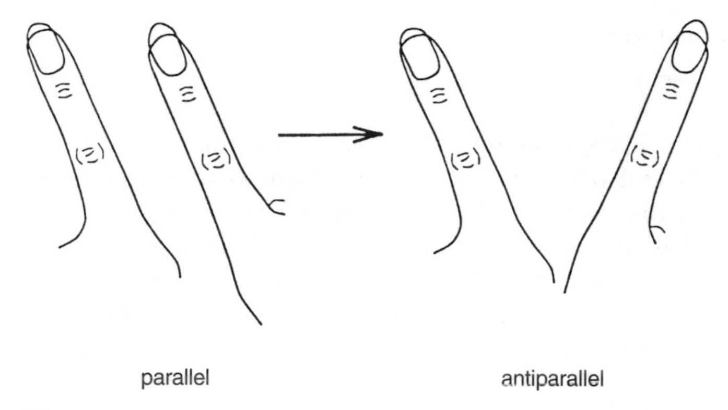

Fig. 9.2. Transition from parallel to symmetric index finger movement

We describe the displacement x_j of index finger $j, j = 1, 2$, (Fig. 9.3) by means of a phase variable ϕ_j

$$x_j = r_j \sin \phi_j \,, \quad j = 1, 2 \,. \tag{9.41}$$

The model, in its most simple form, assumes that the relative phase

$$\phi = \phi_2 - \phi_1 \tag{9.42}$$

obeys the following equation

$$\dot{\phi} = -a \sin \phi - b \sin 2\phi \,. \tag{9.43}$$

By analogy with the mechanical example of Sect. 5.3, ((5.24) and (5.25)), we identify ϕ with the coordinate of a particle that is subjected to a force represented by the r.h.s. of (9.43). Introducing the potential function

$$V(\phi) = -a \cos \phi - \frac{1}{2} b \cos 2\phi \,, \tag{9.44}$$

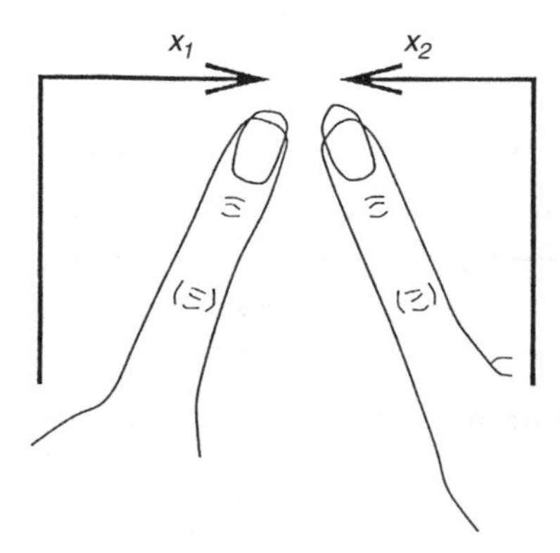

Fig. 9.3. Definition of displacements x_j

we cast (9.43) into the form

$$\dot{\phi} = -\partial V/\partial\phi . \tag{9.45}$$

To fully exploit the analogy between the "movement" of ϕ and that of a particle, we treat the case of overdamped motion. Here we may neglect the term $m\ddot{x}$ in (5.24), so that (5.24) with (5.25) coincides with (9.45), where, of course, $x \leftrightarrow \phi$. Thus, the movement of ϕ corresponds to that of a ball (or particle) in a hilly landscape described by $V(\phi)$. A further crucial idea is now lent from synergetics. It is assumed that the prescribed finger movement frequency ω acts as a *control parameter*. When such a parameter is changed, the corresponding potential "landscape" is deformed. In the present case it is assumed that the ratio b/a in (9.44) depends on ω in such a way that the sequence of landscapes of Fig. 9.4 results. This model can represent (or "explain") or has even predicted a number of experimental findings:

(1) *Hysteresis.* When at low ω the fingers are moved in parallel, the ball is located in the upper minimum ("valley"). With increasing ω, this minimum becomes flatter and flatter and, eventually, disappears. Then the ball rolls down to the deeper valley corresponding to symmetric finger movement. On the other hand, when starting at high ω from this position and ω is lowered, the ball will never "jump up" to its previous higher valley. This means for subjects: To move their fingers in the symmetric mode. Summarizing we may

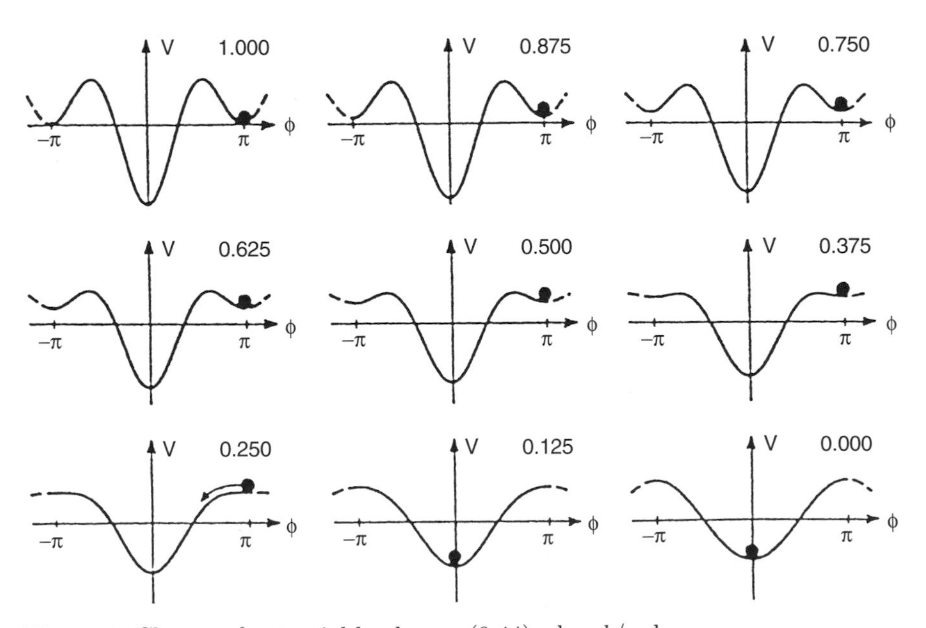

Fig. 9.4. Change of potential landscape (9.44) when b/a decreases

say: For the same frequency ω the finger movement coordination is quite different depending on the *history*.

(2) *Critical slowing down.* Let the prescribed ω be close to its *critical value*, ω_c, where the transition between one kind of movement to the other occurs. When the movement is disturbed, e.g. by small kicks, it takes a comparatively long time until the original, coordinated movement is restored.

(3) *Critical fluctuations.* According to synergetics, close to the critical control parameter values, in the present case ω_c, the variables of a system may undergo pronounced fluctuations. In the present context, such fluctuations of ϕ have been studied by Schöner, Haken and Kelso starting from a Langevin-type equation by including a fluctuating force $F(t)$ in (9.43) (see also Chap. 4). The corresponding experiments by Kelso et al. fully demonstrate the existence of critical fluctuations.

The experimental results (2) and (3) have a fundamental significance for our understanding of the functioning of neural nets. The finger movement is, after all, a manifestation of neural activity. On the other hand, the phenomena (2) and (3) are clear manifestations of self-organization. In other words, we are led to the conclusion that movement coordination is brought about by self-organization and *not* by a motor program in whose frame (2) and (3) would be absent. This point of view is the basis of my book "Principles of Brain Functioning".

9.4 Quadruped Motion

The motion of quadrupeds, such as horses, camels, and so on, has also been modelled by means of phase oscillators, which are assumed to be located in the spinal cord. A remarkable feature of quadruped gaits is that there is quite a variety of them, where walking, trotting and galloping are just a few examples. Furthermore, spontaneous transitions between such gaits may occur.

In the following we wish to write down the basic equations and quote some of the more interesting results. We will assume that all frequencies ω are equal and put

$$\Theta(t) = \omega t + \phi . \tag{9.46}$$

Furthermore we will introduce indices according to the scheme

$$\text{right(r), left(l), front(f), hind(h) limbs} , \tag{9.47}$$

and denote the individual phases correspondingly

$$\phi_{rf}, \phi_{lf}, \phi_{rh}, \phi_{lh} . \tag{9.48}$$

Thus, for instance, ϕ_{rf} is the phase of the right front limb. For later purposes, we will introduce the abbreviation

$$\phi_{ij}, \quad i = r, l; \quad j = f, h, \tag{9.49}$$

where the first index refers to right or left and the second to front or hind. By analogy with what we have seen before in modelling lamprey motion, it suggests itself to introduce the following phase-coupling equations

$$\dot{\phi}_{rf} = a \sin(\phi_{rh} - \phi_{rf}) + b \sin(\phi_{lf} - \phi_{rf}) + c \sin(\phi_{lh} - \phi_{rf}), \tag{9.50}$$

where we have written down just one equation for the right front limb as an example. Clearly, three other equations of the same type have to be added so that one should deal with equations that contain 12 coefficients. As was shown by Schoener et al., because of symmetry considerations, the number of coefficients can be reduced quite appreciably. In order to write down the resulting equations in a concise way, we introduce a "hat notation" by means of

$$\begin{aligned} &\hat{\imath}: \quad \text{if} \begin{cases} i = r, \\ i = l, \end{cases} \text{then} \begin{matrix} \hat{\imath} = l \\ \hat{\imath} = r \end{matrix} \\ &\hat{\jmath}: \quad \text{if} \begin{cases} j = f, \\ j = h, \end{cases} \text{then} \begin{matrix} \hat{\jmath} = h \\ \hat{\jmath} = f. \end{matrix} \end{aligned} \tag{9.51}$$

When diagonal couplings between the limbs are neglected, because of symmetry only two coefficients are needed and the equations for the four limbs can be written in the very concise form

$$\dot{\phi}_{ij} = A_1 \sin\left[\phi_{ij} - \phi_{\hat{\imath}j}\right] + C_1 \sin\left[\phi_{ij} - \phi_{i\hat{\jmath}}\right]. \tag{9.52}$$

Here it is assumed that the symmetry holds with respect to planes through the body of the animal in longitudinal and transverse directions. In order to take care of transitions between gaits in an appropriate fashion, the formulation (9.52) is not sufficient, however, but must be supplemented with the succeeding harmonic terms on the r.h.s. of (9.52) by analogy with procedures introduced by Haken, Kelso and Bunz (see Sect. 9.3) to model changes in finger movement patterns. The resulting equations then acquire the form

$$\begin{aligned} \dot{\phi}_{ij} = &A_1 \sin\left[\phi_{ij} - \phi_{\hat{\imath}j}\right] + A_2 \sin\left[2(\phi_{ij} - \phi_{\hat{\imath}j})\right] \\ &+ C_1 \sin\left[\phi_{ij} - \phi_{i\hat{\jmath}}\right] + C_2 \sin\left[2(\phi_{ij} - \phi_{i\hat{\jmath}})\right]. \end{aligned} \tag{9.53}$$

Many gaits, i.e. movement patterns, have been studied by Schoener et al., where depending on the numerical values of the coefficients the stability regions could be identified. Since the whole procedure has been described in my book "Principles of Brain Functioning", I will not dwell here longer on this problem.

9.5 Populations of Neural Phase Oscillators

The interactions between sets of neural phase oscillators have been studied by numerous authors, who treated various aspects. My following outline is based on a recent book by Tass (see references), who had medical applications, in particular for therapy for Parkinson's disease, in mind. The effect of phase couplings on phase oscillators had been extensively treated by Kuramoto, mainly in the context of chemical waves, but also in the context of neural populations.

9.5.1 Synchronization Patterns

We denote the time-dependent phase of neuron j with eigenfrequency ω_j by ψ_j. The equations are assumed to be in the form

$$\dot{\psi}_j = \omega_j + \frac{1}{N} \sum_{k-1}^{N} M(\psi_j - \psi_k) \, , \quad j = 1, ..., N \, . \tag{9.54}$$

In this approach the function M that provides the coupling between the phases ψ_k and ψ_j does not depend on the indices j and k explicitly. This means that the coupling between different neurons is assumed all the same, an approach called "mean field approximation". M is a 2π-periodic function, and is approximated by the first four Fourier coefficients

$$M(\psi) = - \sum_{m=1}^{4} (K_m \sin(m\psi) + C_m \cos(m\psi)) \, , \tag{9.55}$$

where the use of the sin and cos functions is equivalent to taking into account a phase shift δ that may depend on the index m. An interesting result is the fact that, depending on the values of the parameters ω_j, and K and C, frequency locking between the oscillators becomes possible, i.e. where all oscillators run at the same synchronization frequency

$$\dot{\psi}_j = \Omega^* \quad \text{synchronization frequency} \, , \tag{9.56}$$

but may possess constant phase differences. In this way, one cluster of neurons may become phase-locked, but Tass as well as others studied also the occurrence of up to four clusters and also clustering in the presence of noise.

9.5.2 Pulse Stimulation

Once the possible clusters have been determined, it becomes possible to study the impact of pulsatile stimulation on noisy cluster states. In the context of phase resetting in biology, pulses can – but need not – be short compared to the period of an oscillation. In the work of Tass, the stimulus is described in the form

$$S(\psi) = \sum_{m=1}^{4} I_m \cos(m_\psi + \gamma_m) \,. \tag{9.57}$$

After some time with no stimulus, where the clusters have been formed, the stimulus (9.57) is administered and we study how different types of stimuli act on different types of cluster states. This may, in particular, lead to stimulation-induced desynchronization and spontaneous resynchronization.

More recently, Tass studied the effect of two subsequent stimulations that seem to be more promising for medical applications, e.g. for a treatment of Parkinson's disease.

9.5.3 Periodic Stimulation

The action of a periodic stimulus on a nonlinear oscillator has been studied in many fields of science, and particularly mathematics, under the title "Forced Nonlinear Oscillators". An interesting new feature is the fact that the oscillations can become phase-locked to the phase of the stimulus even if both frequencies are equal or nearly equal, but also if they are in ratio, such as 1:2, or more generally $N : M$, where N and M are integers. Thus, for instance, when the stimulation runs at a frequency $N\Omega$, the resulting oscillation of the system occurs at a frequency $M\Omega$ and the corresponding phases are locked. The periodic stimulation may serve different purposes, for instance to annihilate rhythmic synchronized activity, or to modify ongoing rhythms using a change of frequency, or by reshaping synchronization patterns. In this way, these mathematical studies are of interest to the therapy of Parkinsonian resting tremor.

10. Pulse-Averaged Equations

10.1 Survey

In Sect. 8.13 I have shown how by suitable averages over the equations of a pulse-coupled neural network, equations for the pulse rates and dendritic currents can be obtained (see (8.139) and (8.140)). Equations of a similar type have been derived by Nunez, though along different lines, and we will discuss them below. By the elimination of the axonal pulse rates, we may derive equations for the dendritic currents alone. Since dendritic currents lie at the basis of the formation of electric and magnetic fields of the brain, such equations are of particular interest. We will discuss these and related equations later in this chapter. Here we wish to begin with a comparatively simple case, which has a long history and at the same time reveals highly interesting features of such networks. In the present case, we use equations in which the dendritic currents have been eliminated from (8.139) and (8.140). We then essentially obtain equations that were derived by Wilson and Cowan who called them cortical dynamics equations. Wilson and Cowan established their equations directly, i.e. without an averaging procedure as outlined in Sect. 8.13. They consider two types of neurons, namely
(a) excitatory neurons, whose axonal pulse rate is denoted by E. These are usually pyramidal cells that provide cortical output, and
(b) inhibitory interneurons with axonal pulse rates I. These neurons have usually axons that remain within a given cortical area. Among these neurons, usually all interconnections occur, i.e. $E \to E, E \to I, I \to E, I \to I$. Consistent with anatomy, the recurrent excitations remain relatively localized, whereas inhibitory interactions extend over a broader range (for more recent approaches see the references). In order to make contact with our previous notation in Chap. 8, we note that the pulse rate ω is now interpreted either as E or I

$$\omega \begin{array}{l} \nearrow E \\ \searrow I \,. \end{array} \tag{10.1}$$

Furthermore, we interpret the discrete index j as the position of the corresponding neuron

$$j \to \mathbf{x} \,. \tag{10.2}$$

In the following we will ignore time delays.

10.2 The Wilson–Cowan Equations

Denoting the relaxation time of the axonal pulse rates by τ, and using suitably scaled saturation functions S, the Wilson–Cowan equations can be written in the form

$$\tau \frac{dE(\mathbf{x})}{dt} + E(\mathbf{x}) = S_E\Big(\sum_{\mathbf{y}} w_{EE}(\mathbf{x},\mathbf{y})E(\mathbf{y}) - \sum_{\mathbf{y}} w_{IE}(\mathbf{x},\mathbf{y})I(\mathbf{y}) + P(\mathbf{x})\Big) \tag{10.3}$$

and

$$\tau \frac{dI(\mathbf{x})}{dt} + I(\mathbf{x}) = S_I\Big(\sum_{\mathbf{y}} w_{EI}(\mathbf{x},\mathbf{y})E(\mathbf{y}) - \sum_{\mathbf{y}} w_{II}(\mathbf{x},\mathbf{y})I(\mathbf{y}) + Q(\mathbf{x})\Big) . \tag{10.4}$$

P and Q are external inputs. These authors use the Naka–Rushton function (see Sect. 2.5) in the form

$$S(X) = \begin{cases} \frac{rX^N}{\Theta^N + X^N} & \text{for } X \geq 0 \\ 0 & \text{for } X < 0 \end{cases} \tag{10.5}$$

for the spike rate and make the specific choices

$$N = 2, \; M = 100 \tag{10.6}$$

so that

$$S(X) = \frac{100X^2}{\Theta^2 + X^2} , \tag{10.7}$$

where θ is chosen differently for excitatory and inhibitory neurons, as is indicated by the corresponding indices of S in (10.3) and (10.4). The coupling between the neurons is assumed to be instantaneous, but depends on the distance between the neurons so that we write

$$w_{ij}(\mathbf{x},\mathbf{y}) = w_{ij}(\mathbf{x}-\mathbf{y}) = b_{ij}\exp(- \mid \mathbf{x}-\mathbf{y} \mid /\sigma_{ij}) . \tag{10.8}$$

Since the parameters b and σ are not very accessible experimentally, the choice of reasonable values of parameters rests on some requirements, namely:

1) The resting state $E = 0, I = 0$ will be asymptotically stable in the absence of external inputs P, Q. Because of the form of S, this condition is automatically fulfilled, because the linearization of equations (10.3) and (10.4) yields

$$dE/dt = -\frac{1}{\tau}E, \; dI/dt = -\frac{1}{\tau}I . \tag{10.9}$$

2) The spread of the recurrent inhibition will be larger than that of the recurrent excitation; this is fulfilled by the choice

$$\sigma_{EI} = \sigma_{IE} > \sigma_{EE} \tag{10.10}$$

in (10.8).

3) If no external stimulation is present, no spatially uniform steady state will be possible except the resting state. This sets specific limits on the relations $b_{ij} \cdot \sigma_{ij}$.

Before we proceed to discuss some solutions to the (10.3) and (10.4), we illuminate the properties of the sigmoid function S, which will lead us to some surprises about the properties of the equations of the type of (10.3) and (10.4).

10.3 A Simple Example

Let us consider only one type of neuron E and assume that E is constant in space. Thus we are treating (10.3) with $I = 0$. By a suitable scaling of E, P and time t (see Exercise 1), this equation can be cast into the form

$$\frac{dE}{dt} = -\gamma E + \frac{(E+P)^2}{1+(E+P)^2} \equiv -\gamma E + S(E)\,. \tag{10.11}$$

In the steady state, we have

$$\frac{dE}{dt} = 0\,. \tag{10.12}$$

We first consider no external input, i.e. $P = 0$. Equation (10.11) then acquires the form

$$\gamma E_0(1 + E_0^2) - E_0^2 = 0\,, \tag{10.13}$$

which, quite evidently, has as one solution

$$E_0 = 0\,. \tag{10.14}$$

The remaining equation

$$E_0^2 - E_0/\gamma + 1 = 0 \tag{10.15}$$

possesses two roots

$$E_0 = \frac{1}{2\gamma} \pm \sqrt{1/(2\gamma)^2 - 1}\,. \tag{10.16}$$

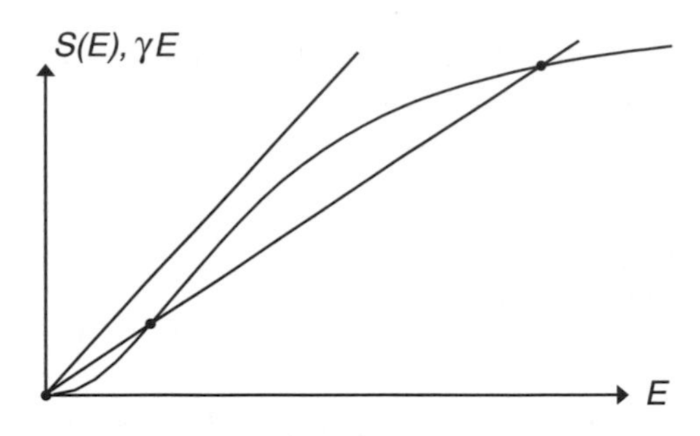

Fig. 10.1. The plot of $S(E)$ (10.11) and γE versus E allows us to determine the zeros of (10.11) for various parameter values

We then readily observe:

$$\begin{aligned}&\text{(a) no real root for} \quad 1/(2\gamma) < 1 \quad \text{(large damping)};\\ &\text{(b) one double root for} \quad 1/(2\gamma) = 1;\\ &\text{(c) two real roots for} \quad 1/(2\gamma) > 1 \quad \text{(small damping)}.\end{aligned} \tag{10.17}$$

Since case (b) is contained in (c) as a special case, we focus our attention on case (c). In this case, we may approximate (10.16) using

$$E_0 \approx \frac{1}{2\gamma}(1 \pm (1 - 2\gamma^2)) = \begin{matrix} \nearrow 1/\gamma - \gamma \approx 1/\gamma \\ \searrow \gamma . \end{matrix} \tag{10.18}$$

These results are illustrated by Fig. 10.1.

An important question is whether the states (10.14) and (10.18) are stable. To check the stability, we write (10.11) in the form

$$\frac{dE}{dt} = -\gamma E + S(E) \equiv f(E), \tag{10.19}$$

where for steady states

$$f(E_0) = 0. \tag{10.20}$$

Putting

$$E = E_0 + \epsilon, \tag{10.21}$$

and using (10.20), we transform (10.19) into

$$\frac{d\epsilon}{dt} = f(E) - f(E_0) \approx \alpha\epsilon, \tag{10.22}$$

where the last term is a good approximation provided f is evaluated for values of E close to one of the steady states. The function $f(E)$ is plotted against E in Fig. 10.2. This figure allows us immediately to derive the stability properties of the individual states as discussed in the figure. The stability of $E_0 = 0$ is no surprise. However, surprisingly enough, there exists also a stable state for $E_0 \neq 0$ even in the absence of any external input. In order to avoid such states, inhibitory neurons are needed. We note already here that highly

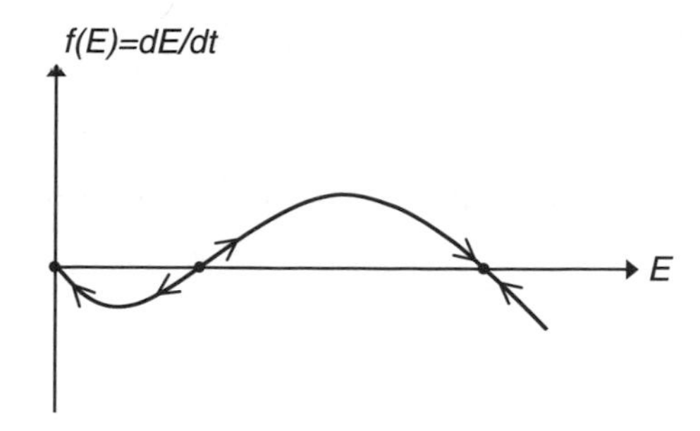

Fig. 10.2. The plot of $f(E) = dE/dt$ allows us to determine the stability of the fixed points. For $f(E) < 0$, the variable E tends to smaller values, and for $f(E) > 0$, E tends to larger values as indicated by the arrows. If the *arrows* aim towards a fixed point, it is stable; unstable otherwise

excited networks can be used as models for epileptic seizures as well as for visual hallucinations which we will discuss below.

Let us now turn to the case with non-vanishing external signal, $P \neq 0$. We introduce a new variable E' by means of

$$E + P = E' \tag{10.23}$$

and consider the steady state. The values of the steady-state solutions can again be determined by graphical representations (cf. Figs. 10.3 and 10.4). Here we wish to present an analytic approach. By means of (10.23), the original (10.11) with (10.12) is transformed into

$$\gamma(E' - P)(1 + E'^2) - E'^2 = 0\,. \tag{10.24}$$

Figure 10.3 suggests that for small enough P one solution, if any, of (10.24) will be a small quantity so that we retain only powers up to second order in E'. Equation (10.24) can then be transformed into

$$E'^2 - \frac{\gamma}{1 + \gamma P} E' + \frac{P\gamma}{1 + \gamma P} = \gamma\,, \tag{10.25}$$

which possesses the roots

$$E' = \frac{\gamma}{2(1 + \gamma P)} \left(1 \pm \sqrt{1 - \frac{4P(1 + \gamma P)}{\gamma}} \right)\,. \tag{10.26}$$

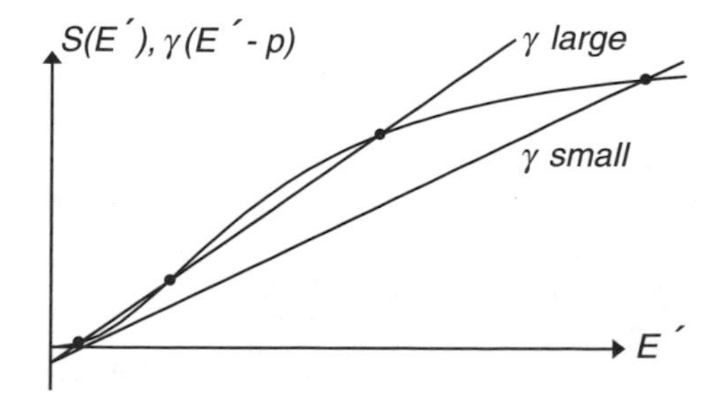

Fig. 10.3. Same as Fig. 10.1, but for $S(E')$ and $\gamma(E' - P)$. Note the possibility of three intersections

We discuss two special cases:

$$(1) \quad \gamma \text{ small and/or } P \text{ large.} \tag{10.27}$$

In this case, no real root exists, at least none that is small in accordance with Fig. 10.3.

Let us consider

$$(2) \quad \gamma \text{ large } (P \text{ small or moderate}). \tag{10.28}$$

In this case, (10.26) can be approximated by

$$E' \approx \begin{matrix} \nearrow (\gamma - P(1+\gamma P))/(1+\gamma P) \\ \searrow P\,, \end{matrix} \tag{10.29}$$

where only the lower root is of interest to us, because it agrees with our original assumption that E' is small.

Let us now consider the upper cross-section in Fig. 10.3. In this case, we are in a region called "saturation". In our analytical approach we assume

$$E' \gg 1 \tag{10.30}$$

and in order to use small quantities, we introduce the inverse of E' via

$$\frac{1}{E'} = \epsilon \ll 1 \tag{10.31}$$

as a new variable. Equation (10.24) can be transformed into

$$\gamma - \gamma\epsilon P + \gamma\epsilon^2 - \gamma\epsilon^3 P - \epsilon = 0\,, \tag{10.32}$$

and ignoring the cubic term of ϵ into

$$\epsilon^2 - \epsilon(1+\gamma P)/\gamma + 1 = 0\,, \tag{10.33}$$

which possesses the roots

$$\epsilon = \frac{1}{2\gamma}(1+\gamma P)\left(1 \pm \sqrt{1 - \frac{(2\gamma)^2}{(1+\gamma P)^2}}\right)\,. \tag{10.34}$$

Again we distinguish between γ large and γ small. In the first case, we obtain

$$\gamma \to \infty,\ \gamma P \text{ finite } \surd \to i\,, \tag{10.35}$$

i.e. no real root exists, at least none that is consistent with (10.30) or (10.31). If, however, γ is small, we obtain approximately

$$\epsilon = \begin{matrix} \nearrow (1+\gamma P)/\gamma - \gamma/(1+\gamma P)^2 \\ \searrow \gamma/(1+\gamma P)\,. \end{matrix} \tag{10.36}$$

Keeping the leading terms and using (10.31), we finally obtain

$$E' = \frac{1}{\epsilon} \approx \begin{cases} \gamma/(1+\gamma P) \\ (1+\gamma P)/\gamma\,. \end{cases} \tag{10.37}$$

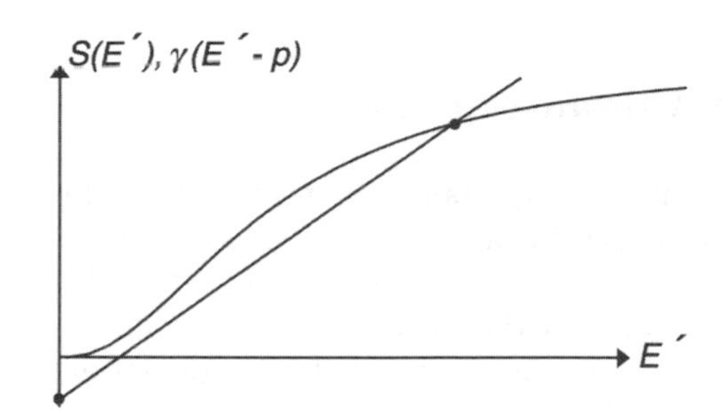

Fig. 10.4. Same as Fig. 10.2, case of saturation

The first root is in contradiction with our assumption (10.30) so that the useful root for us is the second in (10.37). Because of (10.23), we finally obtain

$$E \approx \frac{1}{\gamma} . \tag{10.38}$$

In leading order, E is thus independent of the external signal P. This is typical for *saturation effects* (see Fig. 10.4)

Let us summarize the salient results of this section. The steady states E_0 depend decisively on the size of the damping constant γ (in reduced units). Even in the absence of an external signal, a stable highly excited state may exist. In between the two stable states, there is an unstable steady state that was not shown here. It persists also for non-vanishing external inputs P.

Let us now return to some typical solutions of the Wilson–Cowan equations.

Exercise (Rescaling of (10.3)). The original equation reads

$$\tau \frac{dE}{dt} = -E + \frac{rE^2}{\Theta^2 + E^2} . \quad (^*)$$

We introduce

$$E = \tilde{E}\Theta ,$$

which casts (*) into

$$\tau\Theta \frac{d\tilde{E}}{dt} = -\Theta\tilde{E} + \frac{r\tilde{E}^2}{1+\tilde{E}^2} .$$

Division by r leads to

$$\frac{\tau\Theta}{r}\frac{d\tilde{E}}{dt} = -\frac{\Theta}{r}\tilde{E} + \frac{\tilde{E}^2}{1+\tilde{E}^2} ,$$

which can be cast into the form (10.11) by the choices

$$\frac{r}{\tau\Theta} t = \tilde{t}, \ \frac{\Theta}{r} = \gamma .$$

10.4 Cortical Dynamics Described by Wilson–Cowan Equations

Here we wish to consider the following behavior as an example. We assume that a spatially uniform steady state has been produced by constant input values P, Q. In order to study the stability of this steady state, we linearize the corresponding equations around that state. In this specific example, Wilson and Cowan obtained the following equations for the deviations $E'(x,t), I'(x,t)$ from the steady states

$$\frac{dE'}{dt} = -E' + \frac{1}{4}\int_{-\infty}^{\infty} \exp(-\mid x - x' \mid)E'dx' - \frac{1}{4}\int_{-\infty}^{\infty} \exp(-\mid x - x' \mid /8)I'dx' , \tag{10.39}$$

$$\frac{dI'}{dt} = -I' + \frac{1}{8}\int_{-\infty}^{\infty} \exp(-\mid x - x' \mid)E'dx' - \frac{2}{3}\int_{-\infty}^{\infty} \exp(-\mid x - x' \mid /3)I'dx' , \tag{10.40}$$

where specific numerical values of the constants have been chosen. The authors study specific kinds of spatial dependence, i.e. they put

$$E'(x,t) = E'(t)\cos(kx),\ I'(x,t) = I'(t)\cos(kx) . \tag{10.41}$$

The integrals which occur in (10.39) and (10.40) can be easily evaluated and yield

$$\int_{-\infty}^{+\infty} \exp(-\mid x - x' \mid /\sigma)\cos(kx')dx' = \frac{2\sigma\cos kx}{1+(\sigma k)^2} . \tag{10.42}$$

In this way, (10.39) and (10.40) are transformed into equations for the time-dependent amplitudes E', I'

$$\frac{dE'}{dt} = -E' + \frac{2E'}{1+(4k)^2} - \frac{4I'}{1+(8k)^2} , \tag{10.43}$$

$$\frac{dI'}{dt} = -I' + \frac{2E'}{1+(8k)^2} - \frac{4I}{1+(3k)^2} . \tag{10.44}$$

These two linear equations can be solved using the hypothesis

$$E'(t) = E'_0 e^{\lambda t},\ I'(t) = I'_0 e^{\lambda t} , \tag{10.45}$$

which leads to the eigenvalue equations

$$\lambda E' = c_1 E' - c_2 I' \tag{10.46}$$

and

$$\lambda I' = c_3 E' - c_4 I' , \tag{10.47}$$

where the coefficients are given by

$$c_1 = -1 + \frac{2}{1+(4k)^2}, \; c_2 = \frac{4}{1+(8k)^2}, \tag{10.48}$$

and

$$c_2 = \frac{2}{1+(8k)^2}, \; c_4 = 1 + \frac{4}{1+(3k)^2}. \tag{10.49}$$

As usual, the eigenvalues λ are determined by

$$\begin{vmatrix} c_1 - \lambda - c_2 & \\ c_2 & - c_4 - \lambda \end{vmatrix} = 0 . \tag{10.50}$$

Quite evidently, λ becomes a function of k, i.e. of the spatial variation (10.41)

$$\lambda = \lambda(k) . \tag{10.51}$$

Two limiting cases can be readily obtained, namely

$$k = 0, \quad \lambda = -1, -3 \tag{10.52}$$

and

$$k \to \infty, \quad \lambda = -1, -1 , \tag{10.53}$$

i.e. the steady states are stable. However, in the range $0.085 < k < 0.22$ one value of λ becomes positive, indicating an instability. The authors establish a relationship between this result and the human visual system that is most sensitive to intermediate spatial frequencies. In the numerical experiment, the authors chose a constant spatial input and determined the corresponding spatially uniform resting state that is, of course, non-vanishing now. Then they applied short pulses with a spatial distribution proportional to $\cos(kx)$ that were added to P. For small k, the perturbations had no effect and the system decayed to the uniform equilibrium state and the same was true for large k. For some intermediate values of k, however, even the tiny perturbations led the system to explode into a spatially inhomogeneous state that was asymptotically stable. A typical example of this numerical experiment is shown in Fig. 10.5 from Wilson. An interesting feature of this experiment was that the perturbation giving rise to the spatially structured state had been gone long before the instability erupted. It is this spatially structured instability that gives rise to what is termed a "short-term memory". These authors as well as others find several kinds of behavior that quite often depend critically on parameter values and the initial preparations. Here we quote the following results:

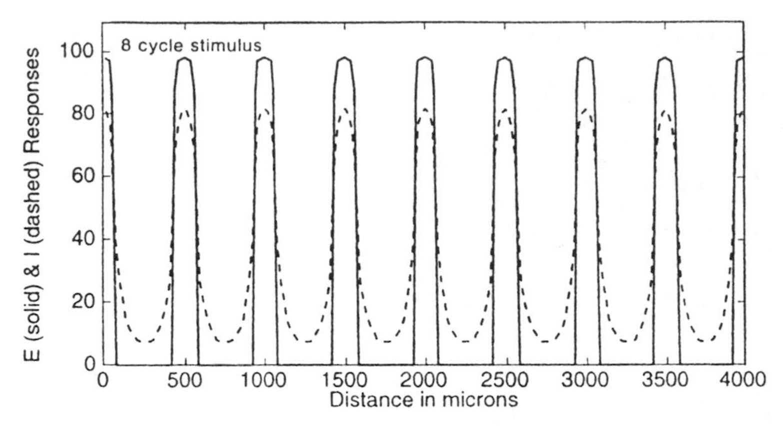

Fig. 10.5. Spatially inhomogeneous steady state triggered by a brief stimulus $0.01\cos(16\pi x)$ (after Wilson, 1999)

1. "Active transient mode". A brief spatial stimulus to a restricted area of the network gives rise to a delayed, but very large, amplification of the network response. This effect is explained by the authors as follows: A local recurrent excitation in the network causes excitatory activity to flare up, but this in turn triggers a delayed inhibitory pulse that subsequently extinguishes the network activity.
2. "Spatially localized oscillatory activity". This activity is produced by a spatial stimulus of some small width. Each peak of this oscillation corresponds to a burst of action potentials. More complex oscillations can be produced by stimuli of greater width.
3. "Travelling waves in response to brief local stimulation". In this case the same parameters as in the oscillatory mode (2) are used, but with a strongly reduced inhibition in effectiveness by providing a constant inhibitory input to the I cells. Here a pair of travelling waves is generated that are running apart in response to a brief stimulus to the center of the network. These waves originate because the reduced inhibitory activity is not sufficient to counteract the activity generated by recurrent activation. These travelling waves are followed by a refractory area, where the inhibition has built up. Because of this trailing refractory zone, two waves that meet will annihilate one another. The authors argue that travelling waves such as these may occur in epileptic seizures.

10.5 Visual Hallucinations

Though in this book I carefully avoid telling stories, I cannot help starting this section with the following reminiscence of mine. During the 1970s and 1980s I organized a series of meetings on synergetics that brought together scientists from quite different disciplines. The neuromathematician Jack Cowan attended such a meeting, where he heard of the spontaneous stripe formation in fluids under specific non-equilibrium conditions. This triggered in him an ingenious idea. He remembered that there is a special mapping from the retina to the visual cortex that can be described by a logarithmic map. Such a map transforms several stripes depending on their direction into concentric circles, radial spokes, or spirals. In fact, a variety of drugs can induce hallucinations involving such patterns that are quite stereotypical and to which also checkerboard hallucinations can be added. Thus, according to Ermentraut and Cowan (1979), when a brain is strongly excited, or, in other words, destabilized, neural activity can develop into roughly parallel stripes. Because of the mapping, the subject will perceive them as concentric circles, radial spokes, or spirals. These authors could show that such stripes can be generated by two-dimensional Wilson–Cowan equations by analogy with the patterns we have got to know above in one dimension. Again by analogy with fluid dynamics, where also square patterns are observed, the Wilson–Cowan network can produce such brain activity patterns.

More recent contributions to such patterns of hallucinations have been made in the theoretical work of Tass. Ermentraut provides a detailed summary of the theoretical work on this fascinating phenomenon.

10.6 Jirsa–Haken–Nunez Equations

These equations belong to the class of phase-averaged equations, but in contrast to the Wilson–Cowan equations they take into account the dynamics of axons as well as of dendrites. While Jirsa and Haken interpret the dendritic activity as done throughout my book, Nunez interprets the corresponding quantities as synaptic activities. Since the dendrites carry charges and currents, they are responsible for the generation of the electric and magnetic fields in the brain. Therefore, it is desirable to derive explicit equations for the dynamics of the dendritic currents. The equations derived by Jirsa and Haken on the one hand and Nunez on the other are quite similar, but differ somewhat in the nonlinearities. In the following we will present the Jirsa and Haken version. In the following we will assume that excitatory neurons have only excitatory synapses and inhibitory neurons only inhibitory synapses. Furthermore, we assume that in ensembles of neurons the conversion of axonal pulses into dendritic currents at the synapses is linear, whereas the conversion of dendritic currents into axonal pulses at the somas is sigmoidal, i.e. nonlinear. The spatial distribution range of the dendrites and intracortical fibres is assumed to be very short, only the cortical connections may cause a significant delay via propagation. It is assumed that this delay can be described by means of a constant propagation velocity and the distance. Finally, the external input is modelled so that afferent fibers make synaptic connections. We will distinguish between excitatory and inhibitory dendritic currents ψ using the indices e and i, respectively, and between the pulse rates of excitatory and inhibitory neurons using E and I, respectively. The connection between the axonal pulse rates and the dendritic currents can then be written in the form

$$\psi_e(x,t) = a_e \int\limits_{\Gamma} f_e(x,x')E(x',t- \mid x-x' \mid /v_e)dx' \tag{10.54}$$

and

$$\psi_i(x,t) = a_i I(x,t) \,, \tag{10.55}$$

where a_e, a_i are the corresponding synaptic strengths and f_e is a function that decays with distance

$$f_e(x,x') = \frac{1}{2\sigma} \exp(- \mid x-x' \mid /\sigma) \,. \tag{10.56}$$

v_e is the propagation velocity in the axon of the excitatory neurons. These equations are understood as referring to ensembles of neurons. The integration is extended over the area Γ of the neural sheet that in our model we assume is one-dimensional. The conversion of dendritic currents into axonal pulse rates is described by the equations

$$E(x,t) = S_e\left[\psi_e(x,t) - \psi_i(x,t) + p_e(x,t)\right] \tag{10.57}$$

and

$$I(x,t) = S_i\left[\psi_e(x,t) - \psi_i(x,t) + p_i(x,t)\right] \,, \tag{10.58}$$

where S_e, S_i are the corresponding sigmoidal functions. Because of the linear relationships (10.54) and (10.55), it is a simple matter to eliminate either ψ or E, I.

(1) Elimination of the dendritic currents.
When we insert the expressions (10.54) and (10.55) into the r.h.s. of (10.57) and (10.58), we readily obtain

$$E(x,t) = S_e\left[a_e \int\limits_{\Gamma} f_e(x,x')E(x',t- \mid x-x' \mid /v_e)dx' - a_i I(x,t) + p_e(x,t)\right] \tag{10.59}$$

and

$$I(x,t) = S_i\left[a_e \int\limits_{\Gamma} f_e(x,x')E(x',t- \mid x-x' \mid /v_e)dx' - a_i I(x,t) + p_i(x,t)\right] \,. \tag{10.60}$$

These are essentially the Wilson–Cowan equations in their general form, where delays are taken into account, but where now, in our case, the axonal decay time is assumed to be very short.

(2) Elimination of the axonal pulse rates.
In this case, we just have to insert the relations (10.57) and (10.58) into the r.h.s. of (10.54) and (10.55). This yields

$$\begin{aligned}\psi_e(x,t) = a_e \int_\Gamma f_e(x,x') S_e \Big[&\psi_e(x', t- \mid x-x' \mid /v_e) \\ &-\psi_i(x', t- \mid x-x' \mid /v_e) \\ &+p_e(x', t- \mid x-x' \mid /v_e) \Big] dx' \end{aligned} \tag{10.61}$$

and

$$\psi_i(x,t) = a_i S_i[(\psi_e(x,t) - \psi_i(x,t) + p_i(x,t)] \, . \tag{10.62}$$

Taking into account that the r.h.s. of (10.62) can be well approximated by a linear function, i.e. $S_i(X) \approx \alpha_i X$, we readily obtain

$$\psi_i(x,t) \approx \frac{a_i \alpha_i}{1 + a_i \alpha_i} (\psi_e(x,t) + p_i(x,t)) \, . \tag{10.63}$$

Inserting (10.63) into (10.61), we obtain an equation for the dynamics of the excitatory synaptic activity

$$\begin{aligned}\psi_e(x,t) = a_e \int_\Gamma f(x,x') S_e \Big[&a\psi_e(x', t- \mid x-x' \mid /v_e) \\ &+p(x', t- \mid x-x' \mid /v_e) \Big] dx' \, , \end{aligned} \tag{10.64}$$

where

$$a = 1 - \frac{a_i \alpha_i}{1 + a_i \alpha_i} \, , \tag{10.65}$$

and

$$p(x,t) = p_e(x,t) - \frac{a_i \alpha_i}{1 + a_i \alpha_i} p_i(x,t) \, . \tag{10.66}$$

Since the unknown function ψ_e occurs on the r.h.s. of (10.64) under an integral, we have to deal with an integral equation, which is usually a rather complicated task. However, it is possible to convert this equation into a differential equation. To this end, we convert the integral over the spatial coordinates in (10.64) into an equation over both spatial coordinates and time by means of the δ-function

$$\delta(t - t' - \mid x - x' \mid /v_e) \, . \tag{10.67}$$

This allows us to express the r.h.s. of (10.64) by means of a Green's function in the form

$$\psi_e(x,t) = \int_{\Gamma} \int_{-\infty}^{\infty} G(x-x', t-t')\rho(x',t')dx'dt' , \tag{10.68}$$

where the Green's function is given by

$$G(x-x', t-t') = \delta(t-t'- \mid x-x' \mid /v_e)\frac{1}{2\sigma}e^{-|x-x'|/\sigma} . \tag{10.69}$$

We used the abbreviation

$$\rho(x,t) = S_e \left[a\psi_e(x,t) + p(x,t)\right] . \tag{10.70}$$

To proceed further, we introduce the Fourier transforms

$$\psi_e(x,t) = \frac{1}{(2\pi)^2} \int_{-\infty}^{+\infty}\int_{-\infty}^{+\infty} e^{ikx-i\omega t}\tilde{\psi}_e(k,\omega)dkd\omega , \tag{10.71}$$

$$\rho(x,t) = \frac{1}{(2\pi)^2} \int_{-\infty}^{+\infty}\int_{-\infty}^{+\infty} e^{ikx-i\omega t}\tilde{\rho}(k,\omega)dkd\omega \tag{10.72}$$

and

$$G(\xi,\tau) = \frac{1}{(2\pi)^2} \int_{-\infty}^{+\infty}\int_{-\infty}^{+\infty} e^{ik\xi-i\omega\tau}\tilde{G}(k,\omega)dkd\omega \tag{10.73}$$

with the abbreviation

$$\xi = x - x', \ \tau = t - t' . \tag{10.74}$$

It is well known from the theory of Fourier transforms that a convolution of the form (10.68) transforms into a simple product of the Fourier transforms $\tilde{\psi}_e, \tilde{\rho}, \tilde{G}$,

$$\tilde{\psi}_e(k,\omega) = \tilde{G}(k,\omega)\tilde{\rho}(k,\omega) , \tag{10.75}$$

where

$$\tilde{G}(k,\omega) = \frac{\omega_0^2 - i\omega_0\omega}{v_e^2k^2 - (\omega_0 - i\omega)^2} \tag{10.76}$$

with the abbreviation

$$\omega_0 = \frac{v_e}{\sigma_e} . \tag{10.77}$$

We multiply both sides of (10.75) by the denominator of (10.76). Now again from the theory of Fourier transformation (see also Exercise 1), multiplication of the Fourier transform of a time-dependent function by $i\omega$ corresponds to a time differentiation, whereas multiplication of the Fourier transform of a space-dependent function by k^2 corresponds to the negative second derivative. Thus, when taking the reverse transformation of (10.75) from k, ω space into x, t space, we arrive at the equation

$$\ddot{\psi}_e + \left(\omega_0^2 - v_e^2 \Delta\right) \psi_e + 2\omega_0 \dot{\psi}_e = \left(\omega_0^2 + \omega_0 \frac{\partial}{\partial t}\right) \rho(x, t) . \tag{10.78}$$

This is a typical wave equation, which depends on a source term $\rho(x, t)$. The difficulty in solving (10.78) arises from the fact that ρ depends on the unknown variable ψ_e in a nonlinear fashion. Thus, in order to solve (10.78), one has to resort to numerical integration. In order to obtain a first insight into the solutions of (10.78), some approximations are required that we will discuss in the next section, where contact with experimental results will be made.

Exercise 1. Show that multiplication of the Fourier transform of a time-dependent function by $i\omega$ corresponds to a time differentiation. Multiplication of the Fourier transform of a space-dependent function by k^2 corresponds to the negative second spatial derivative.
Hint: Use the explicit form of the Fourier transform.

10.7 An Application to Movement Control

In this section, I wish to show how the field (or wave) equation (10.78) of the preceding section can be used to bridge the gap between processes at the neuronal level and macroscopic phenomena that are observed at the behavioral level and at the level of MEG (magnetoencephalogram) recordings. This and related phenomena have been described in extenso in my book "Principles of Brain Functioning" as well as in Kelso's book "Dynamic Patterns".

10.7.1 The Kelso Experiment

Kelso et al. exposed a subject to an acoustic signal that was composed of equidistant beeps. The subject had to push a button in between each beep, i.e. in a syncopating manner. The stimulus frequency, i.e. the inverse of the time interval between the beeps, was set to 1 Hz at the beginning and was increased by 0.25 Hz after 10 stimulus repetitions up to 2.25 Hz. Around the frequency of 1.75 Hz, the subject spontaneously and involuntarily switched to a synchronized motion. This switch is termed the "nonequilibrium phase transition". During this experiment, the magnetic field was measured over the left parieto temporal cortex, namely covering the motor and auditory areas.

Detailed analysis conducted by Fuchs et al. revealed that in the pretransition region the registered brain signals oscillate mainly with the stimulus frequency. The stimulus and the motion response frequencies are locked. Then, at the phase transition point, a switch in behavior occurs. The brain signals now oscillate with twice the stimulus frequency in the post-transition region. Applying a Fourier transformation to the signals and looking at the components of the same frequency as the stimulus, one observes that the phase difference between the stimulus signal and the time series is stable in the pretransition region, but undergoes an abrupt change of π at the transition point and remains stable again in the post-transition region. Near the transition point, typical features of non-equilibrium phase transitions like critical slowing down and critical fluctuations (see Sect. 9.3) are observed in both the behavioral data and the brain signals. In the present context we will not be concerned with these phenomena, however. In order to analyze the spatiotemporal patterns before and after the transition, a basic idea of synergetics was invoked, namely that close to transition points the dynamics even of a complex system is governed by a few dynamic variables, the so-called order parameters. Indeed, a detailed analysis performed by Jirsa et al. showed that the spatiotemporal patterns are dominated by two basic spatial patterns $v_0(\mathbf{x})$ and $v_1(\mathbf{x})$ so that the field pattern over the measured SQUID area can be written in the form

$$q(\mathbf{x},t) = \xi_0(t)v_0(\mathbf{x}) + \xi_1(t)v_1(\mathbf{x}) + \quad \text{add. terms}\,, \tag{10.79}$$

where the additional terms contain so-called enslaved modes and represent comparatively small corrections to the first two terms on the r.h.s. of (10.79). The decomposition (10.79) is based on an improvement of the Karhunen–Loève expansion (principal component analysis, singular value decomposition). The observed dynamics of the order parameters ξ_0, ξ_1 and especially its behavior at the transition point was modelled using dynamic equations by Jirsa et al.

$$\begin{aligned} &\underbrace{\ddot{\xi}_0}_{(1)} + \underbrace{\omega_0^2\xi_0}_{(2)} + \underbrace{\gamma_0\dot{\xi}_0}_{(3)} + b_0 \underbrace{\xi_0^2}_{(4)} \dot{\xi}_0 + c_0 \underbrace{\xi_1^2}_{(4a)} \dot{\xi}_0 \\ &+\epsilon_0 \sin(\underbrace{2\Omega}_{(5)} t)\xi_0 + \delta_0 \underbrace{\sin(\Omega\, t)}_{(6)}\xi_1^2 + \underbrace{d_0}_{(8)} \dot{\xi}_1 = 0\,. \end{aligned} \tag{10.80}$$

The equations were formulated in an entirely symmetric fashion so that the equation for ξ_1 can be obtained from (10.80) by exchanging the indices 0 and 1 everywhere

$$\text{exchange indices} \quad 0 \leftrightarrow 1\,. \tag{10.81}$$

The effect of the acoustic signal at frequency Ω enters (10.80) and (10.81) in the form of the terms

$$\sin(2\Omega t)\xi_j, \quad \sin(\Omega t)\xi_j^2\,. \tag{10.82}$$

10.7.3 The Field Equation and Projection onto Modes

We may now insert the expressions for $p_{\text{ext.}}$ with (10.84) and (10.89) into the field equation (10.78). To proceed further, we make two approximations, namely we assume that ψ_m (10.85) is a small quantity so that it is sufficient to retain only the linear term on the r.h.s. of the field equation (10.78). Furthermore, we assume that the acoustic signal is comparatively fast so that

$$\omega_0^2 \psi_m \ll \omega_0 \frac{\partial}{\partial t} \psi_m \tag{10.90}$$

holds. In this way, the field equation can be cast into the form

$$\begin{aligned} &\ddot{\psi}_e + \left(\omega_0^2 - v_e^2 \Delta\right) \psi_e + 2\omega_0 \dot{\psi}_e \\ &= \left(\omega_0^2 + \omega_0 \frac{\partial}{\partial t}\right) S_e \left[\alpha \psi_e(x,t) + p_a(x,t)\right] + \tilde{d}(x) \dot{\psi}_m(t) \,. \end{aligned} \tag{10.91}$$

The function $\tilde{d}(x)$ is proportional to $d(x)$ in (10.89). To proceed further, we approximate the threshold function by means of

$$S_e(X) = \alpha X - \frac{4}{3} \alpha^3 X^3 \,. \tag{10.92}$$

It is now a simple though slightly tedious matter to write down the field equation explicitly, and I leave it as an exercise to the reader to determine the coefficients explicitly

$$\begin{aligned} &\underbrace{\ddot{\psi}_e}_{(1)} + \left(\Omega_0^2 - \underbrace{v_e^2}_{(2)} \Delta \right) \psi_e + \underbrace{\gamma \dot{\psi}_e}_{(3)} + A \underbrace{\psi_e^3}_{(3a)} + B \underbrace{\psi_e^2}_{(4)} \dot{\psi}_e + \underbrace{K_1}_{(5)} + \underbrace{K_2}_{(6)} \\ &\quad + \underbrace{K_3}_{(7)} + \gamma_1 \underbrace{(x)}_{(8)} \dot{\psi}_m = 0 \,. \end{aligned} \tag{10.93}$$

Some of the linear terms in S_e have the same form as terms that are already present on the l.h.s. of (10.91) so that we may take care of them just by new constants, for instance using the replacement

$$\omega_0 \to \Omega_0, \dots \,. \tag{10.94}$$

The terms (5), (6) and (7) represent the effect of the acoustic driving signal on the field ψ_e and are specific linear combinations of the following terms:

$$(5) \quad K_1: \quad \sin(2\Omega t)\psi_e, \quad \cos(2\Omega t)\dot{\psi}_e, \quad \cos(2\Omega t)\psi_e \,; \tag{10.95}$$

$$(6) \quad K_2: \quad \cos(\Omega t)\psi_e^2, \quad \sin(\Omega t)\psi_e\dot{\psi}_e, \quad \sin(\Omega t)\psi_e^2 \,; \tag{10.96}$$

$$(7) \quad K_3: \quad \cos(\Omega t), \quad \sin(\Omega t), \quad \cos(3\Omega t), \quad \sin(3\Omega t) \,. \tag{10.97}$$

Part IV

Conclusion

11. The Single Neuron

11.1 Hodgkin–Huxley Equations

In our book, the study of neural nets composed of many neurons is in the foreground of our interest. To this end, we had to make a number of simplifying assumptions on the behavior of individual neurons. In this chapter we want to get acquainted with more precise models of individual neurons, whereby we will focus our attention on the generation of axonal pulses. This will also allow us to judge the approximations we had to make in our network analysis. The fundamental equations describing the generation of action potentials and their spiking property had been established by Hodgkin and Huxley as early as 1952. In Sect. 11.1 we will discuss these equations and some simplifying approaches to them. In order to elucidate the origin of the spiking behavior of the solutions, we will discuss the FitzHugh–Nagumo equations in Sect. 11.2 and then return in Sect. 11.3 to the Hodgkin–Huxley equations and especially to generalizations of them. The axon membrane is a lipid bilayer that may be considered as a thin insulating sheet that can store electric charges like a conventional capacitor. According to electrostatics, the capacitance C, the voltage V across the membrane and the charge on it are related by

$$CV = Q\,. \tag{11.1}$$

When currents flow between the two sides of the capacitor, as happens in the axon membrane through ion channels, the voltage V changes in the course of time, where the negative time derivative of Q is just the electric current across the membrane

$$C\frac{dV}{dt} = -I\,. \tag{11.2}$$

The current I is composed of current distributions due to different ions Na,K, due to leakage, and due to an input current I_{input}

$$I = I_{\text{Na}} + I_{\text{K}} + I_{\text{leak}} + I_{\text{input}}\,. \tag{11.3}$$

The size of the currents is determined by Ohm's law

$$I = g(V - E)\,, \tag{11.4}$$

where g is the electric conductance and E the equilibrium potential of the ion. E is determined by the Nernst equation that can be derived from thermodynamics

$$E = \frac{RT}{zF} \ln \left(C_{out}/C_{in} \right) . \qquad (11.5)$$

The quantities in (11.5) have the following meaning: C_{out}/C_{in} concentrations of ions outside/inside the cell; R, thermodynamic gas constant; F, Faraday constant; T, absolute temperature; at 20^0 C, $RT/F = 25$ mV; z, valency of an ion.

As a result of their experiments, Hodgkin and Huxley concluded that the conductances are not constant but depend on the membrane potential V and furthermore that they are dynamic quantities that are determined by differential equations of first order. All in all, the Hodgkin–Huxley equations comprise four highly nonlinear differential equations each of first order. In 1985 Rinzel observed that these equations can be reduced to two differential equations each of first order. According to Wilson (1999), they can be further simplified, to an excellent degree of approximation, to the following form

$$C\frac{dV}{dt} = -g_{\mathrm{Na}}(V)(V - E_{\mathrm{Na}}) - R(V - E_{\mathrm{K}}) + I_{\mathrm{input}} \qquad (11.6)$$

and

$$\frac{dR}{dt} = \frac{1}{\tau_R}(-R + G(V)) , \qquad (11.7)$$

where the individual quantities are numerically given by: $C = 0.8\ \mu F/cm^2$; $g_{\mathrm{Na}}(V) = 17.81 + 47.71V + 32.63V^2$; $E_{\mathrm{Na}} = 0.55 \ \hat{=}\ 55$ mV; $E_{\mathrm{K}} = -0.92 \ \hat{=}\ -92$ mV; $\tau_R = 1.9$ ms; $G(V) = 1.03 + 1.35$ V; and R on the r.h.s. of (11.6) is replaced by 26.0 R, where R is again the gas contant.

Note that the timescale is measured in milliseconds and that the actual voltages are divided by a factor of 100 so that for example, the equilibrium potential E_{Na} of 0.55 corresponds to 55 mV, as indicated above.

Looking at the mathematical structure of the r.h.s., we readily note that the r.h.s. of (11.6) is a polynomial in V up to the third order, while the r.h.s. of (11.7) is a linear function both in R and V. But why do (11.6) and (11.7), which describe a fundamental physical process, describe spiking? To this end, let us consider equations derived later by FitzHugh as well as Nagumo that have a structure quite similar to (11.6) and (11.7), though in that case the physicochemical underpinning is not so evident.

11.2 FitzHugh–Nagumo Equations

The FitzHugh–Nagumo equations read

$$\frac{dV}{dt} = \gamma_V \left(V - \frac{V^3}{3} - R + I_{\mathrm{input}} \right) , \quad \gamma_V = 10 , \qquad (11.8)$$

and

$$\frac{dR}{dt} = \gamma_R(-R + 1.25V + 1.5)\,, \quad \gamma_R = 0.8\,, \tag{11.9}$$

where different authors use somewhat different parameters. In the following we will use those of Wilson. If there is no input,

$$I_{\text{input}} = 0\,, \tag{11.10}$$

one may show that there is only one equilibrium point for which

$$\frac{dV}{dt} = 0, \quad \frac{dR}{dt} = 0\,. \tag{11.11}$$

Using (11.11), (11.8) and (11.9), one may readily verify that this point lies at

$$V = -1.5, \quad R = -\frac{3}{8}\,. \tag{11.12}$$

The conventional stability analysis by means of linearization reveals that this point is stable. A key to the understanding of the behavior of the solutions to (11.8) and (11.9) lies in the time constants that have a large ratio

$$\gamma_V/\gamma_R = 12.5\,. \tag{11.13}$$

This implies that V relaxes very quickly, whereas R does so only slowly. In an extreme case that we will discuss first we may assume

$$\frac{dR}{dt} \approx 0\,, \tag{11.14}$$

so that R in (11.8) may be considered as constant. Yet (11.8) remains a non-linear equation, whose solutions seem to be difficult to discuss. But a little trick from physics helps a lot, namely we identify the behavior of V as that of the coordinate x of a particle that is subject to a force and whose damping is so large that we may ignore the acceleration term occurring in Newton's law. Actually, we used the same trick above, e.g. in Sect. 9.1. As is known from mechanics, the force can be written as the negative derivative of a potential function W so that we cast (11.8) into the form

$$\frac{dV}{dt} = -\gamma_V \frac{\partial W}{\partial V}\,. \tag{11.15}$$

As one may readily verify by comparison between (11.15) and (11.8), W is given by

$$W = -\frac{V^2}{2} + \frac{V^4}{12} + (R - I_{\text{input}})\,V\,. \tag{11.16}$$

For $I_{\text{input}} = 0$ and $R = -3/8$ (see (11.12)), W is plotted in Fig. 11.1, where the equilibrium value $V = -1.5$ is indicated. When I_{input} is increased,

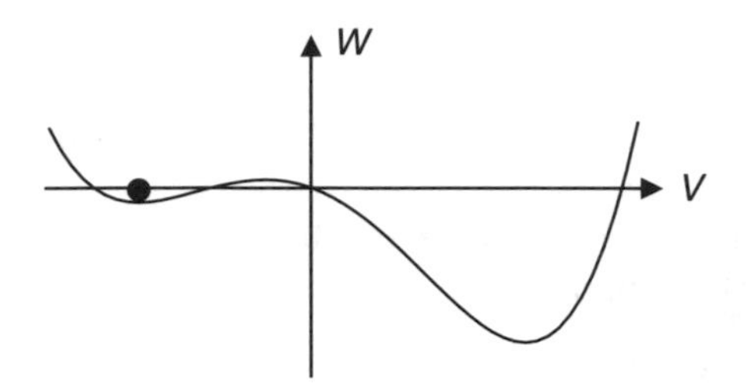

Fig. 11.1. The plot of W (11.16) versus V shows two valleys

the curve $W(V)$ is tilted more and more to the right. When will the particle leave its left position in Fig. 11.1 and slide down to the right minimum? This requires that the slope

$$\frac{dW}{dV} = -V + \frac{V^3}{3} + (R - I_{\text{input}}) \tag{11.17}$$

is negative everywhere on the r.h.s. of the left minimum, at least until the right minimum of W is reached. To this end, let us look at the extrema of dW/dV. This is a condition on the size of $R - I_{\text{input}}$ that is considered as a fixed parameter. Because dW/dV depends also on V, we must proceed in two steps. Since dW/dV must be < 0 in the required interval of V, we first determine the value(s) for which dW/dV reaches its maximum.

This maximum (or extremum) lies at (see Fig. 11.2b)

$$\frac{d^2W}{dV^2} = 0\,, \tag{11.18}$$

from which we deduce

$$V = \pm 1 \tag{11.19}$$

and we find the extremal value

$$\frac{dW}{dV} = \mp\frac{2}{3} + (R - I_{\text{input}}) \quad \text{for} \quad V = \pm 1\,. \tag{11.20}$$

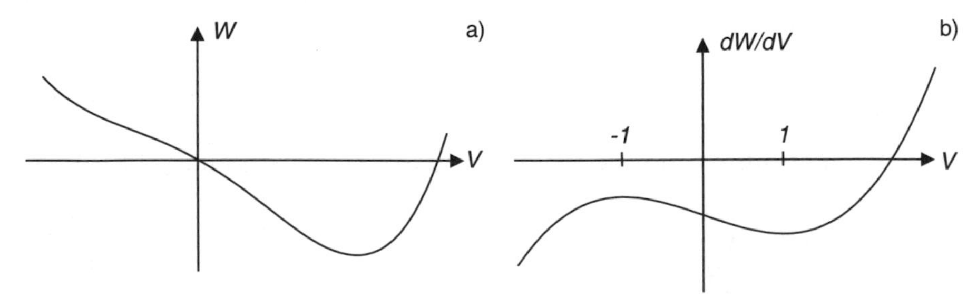

Fig. 11.2. a) For I_{input} large enough, only one valley of $W(V)$ remains. **b**) For $R - I_{\text{input}}$ sufficiently negative, $dW/dV < 0$ for $V < 1$

In the present case, we have to consider

$$V = -1 \,, \quad \frac{dW}{dV} = \frac{2}{3} + (R - I_{\text{input}}) \,. \tag{11.21}$$

The instability of the left position of the particle occurs if

$$\frac{dW}{dV} < 0 \,. \tag{11.22}$$

This implies

$$\frac{2}{3} + R - I_{\text{input}} < 0 \,, \tag{11.23}$$

or using the equilibrium value of R (11.12)

$$R = -\frac{3}{8} \,, \tag{11.24}$$

so that finally

$$I_{\text{input}} > \frac{7}{24} \approx 0.3 \,. \tag{11.25}$$

A comparison with the exact threshold value of I_{input} shows that (11.25) is of the correct order of magnitude. So let us consider a value of I_{input} that is large enough, and take as an example

$$I_{\text{input}} = 1.5 \,. \tag{11.26}$$

In this case, we find only one steady state

$$V = 0 \,, \quad R = 1.5 \tag{11.27}$$

that is, however, unstable. This can be easily seen by looking at Fig. 11.3. Now look what is happening in such a case. While R can be considered as practically constant, V will quickly move to its equilibrium value that is denoted by V_+ in the figure. As a consequence of this, the r.h.s. of (11.9) becomes large so that

$$\frac{dR}{dt} > 0 \,. \tag{11.28}$$

R increases and this implies that, according to Fig. 11.4, the potential curve is tilted to the l.h.s. V will now move very quickly to V_-, in which case the r.h.s. of (11.9) becomes negative and R decreases

$$\frac{dR}{dt} < 0 \,. \tag{11.29}$$

As a more detailed analysis shows, the changes of R need not be large so as to produce the switch from Fig. 11.2a to 11.4 and back again. On the other hand, the differences between V_+ and V_- may be large, i.e. the action potential can show large changes, or, in other words, spikes. Because the changes needed

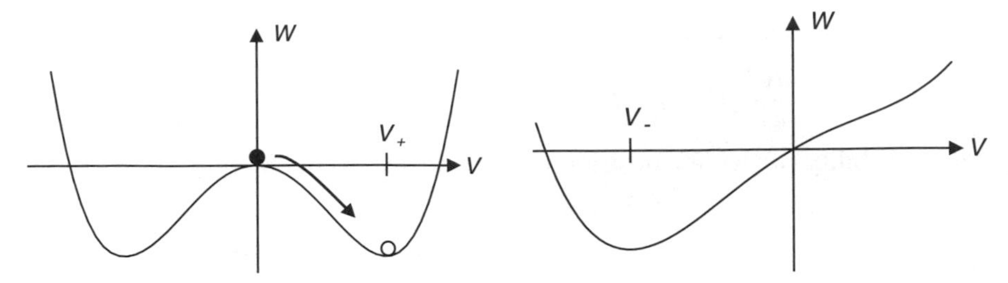

Fig. 11.3. W(11.16) for $R = I_{\text{input}} = 1.5$

Fig. 11.4. W(11.16) for $R \gg I_{\text{input}}$

in R need not be large, in spite of the relatively small constant γ_R, the short time suffices to switch from Fig. 11.2a to Fig. 11.4 and back gain.

It should be noted that these switching times are different, because (11.9) is asymmetric with respect to V because of the additive constant $1.5\gamma_R$. Our above considerations are surely not an exact derivation of the behavior of the solution of the FitzHugh–Nagumo equations, but they may provide the reader with some insight into the origin of that behavior.

11.3 Some Generalizations of the Hodgkin–Huxley Equations

Equations (11.6) and (11.7) provide an accurate and tractible description of the Hodgkin–Huxley equations for action potentials in the squid axon. However, the squid axon is unusual in having only one Na^+ and one K^+ current. As a consequence, the squid axon cannot fire at rates below about 175 spikes/s. It produces only a modest increase in spike rates with increasing input current. The vast majority of neurons in other animals also possess a rapid transient K^+ current that permits the cell to fire at very low spike rates with a long latency to the first spike when the input current is low. This current known as I_{a} was first characterized and added to the Hodgkin–Huxley model by Connor, Walter and McKown (1977). I_{a} currents are found in a wide range of neurons, including human and mammalian neocortical neurons. Mathematically speaking, we must alter the original equations (11.6) and (11.7) in such a way that there is a pronounced asymmetry in the switching between Figs. 11.2a and 11.4. It can be shown that this can be achieved by using a quadratic function G in (11.7). Thus, again following Wilson (1999), the corresponding equations read

$$\frac{dV}{dt} = -\left(17.81 + 47.58V + 33.8V^2\right)(V - 0.48) \\ -26R(V + 0.95) + I \tag{11.30}$$

and

$$\frac{dR}{dt} = \frac{1}{\tau_R}(-R + 1.29V + 0.79 + 0.33(V + 0.38)^2), \quad (11.31)$$

where the constants are given by

$$C = 1.0\,\mu\text{F/cm}^2, \quad \tau_R = 5.6\,\text{ms}. \quad (11.32)$$

We leave it to the reader as an exercise to discuss the effect of the additional quadratic term in (11.31) on switching the size of the r.h.s. of (11.31).

An interesting property of another class of neurons is neural bursting. There is a rapid sequence of spikes that is followed by a refractory period. Then again the same series of spikes is emitted and so on. Do we have any idea how to model that? To this end, consider the role of the input I_{input}. As we have seen before, there is a threshold below which there is no spiking, above which there is spiking. Thus if we let the neuron modulate its effective input, we would obtain just that behavior. In fact, such a behavior can be mimicked by adding a new term to the r.h.s. of (11.30)

$$C\frac{dV}{dt} = \{11.30\} - 0.54H(V + 0.92), \quad (11.33)$$

leaving (11.7) for R unchanged

$$\frac{dR}{dt} = \{11.7\}, \quad (11.34)$$

but introducing a dynamic equation for H in the form

$$\frac{dH}{dt} = \frac{1}{250}(-H + 9.3(V + 0.70)). \quad (11.35)$$

In (11.33), (11.34) we used the abbreviation $\{X\}$ = r.h.s. of Eq. (X). C in (11.33) is chosen to be 0.8 and the numerical factors in (11.33) differ slightly from those of (11.30). Because the relaxation time of H, 250 ms, is very large, H may change very slowly, which results in the desired slow modulation effect of the effective threshold.

11.4 Dynamical Classes of Neurons

According to Connors and Gutnick (1990), Gutnick and Crill (1995), as well as Gray and McCormick (1996) (see also Wilson (1999)), the mammalian cortex contains four classes of neurons that can be differentiated with respect to the spiking responses to sustained intercellular current injection.

1. Fast spiking cells that are described by (11.30) and (11.31) with $\tau_R =$ 2.1 ms.

2. Regular spiking cells. These neurons show spike frequency adaptation, i.e. the spike rate rapidly decreases during continuous stimulation. This effect can be taken care of by adding a further term to (11.30) so that (11.36) results, whereas (11.31) remains unchanged (11.37).

$$\frac{dV}{dt} = \{11.30\} - 13H(V + 0.95)\,, \tag{11.36}$$

$$\frac{dR}{dt} = \{11.31\}\ . \tag{11.37}$$

However, a new dynamic equation for H is required that has the form

$$\frac{dH}{dt} = \frac{1}{99.0}(-H + 11(V + 0.754)(V + 0.69))\,. \tag{11.38}$$

Here we used the abbreviation $\{X\}$ defined above.
3. Neocortical bursting cells.
4. Intrinsic bursting cells. Both (3) and (4) must be described by four first-order differential equations, where different parameters must be used to cover (3) or (4). For the purpose of this book it may may suffice to quote the first model that is due to Chay and Keizer (1983) as shown above in (11.33)–(11.35). The four first-order differential equations are quite similar in structure to these equations.

11.5 Some Conclusions on Network Models

The classification of neurons makes it possible, at least to some extent, to discuss the approximations that we made in devising the model equations for the neural network. Obviously the fast spiking cells, as described by (11.30) and (11.31), obey the model assumptions quite well. The dynamics of bursting cells can be modelled by treating an individual burst as a thick spike that then in some approximation can be represented again as a thin spike provided the relaxation rates of the process are large compared to the spike width. It should be not too hard to mimic also spike frequency adaptation, though this is beyond the scope of the present book.

12. Conclusion and Outlook

An obvious question for the conclusion may be: What did the reader learn from this book? At the core of it there were two approaches to pulse-coupled neural nets: the lighthouse model and integrate and fire models. I tried a systematic exposition, which made all calculations explicit. So I am sure the reader could follow step by step. Let us first recapitulate our results on the *lighthouse model*. We established the phase-locked state, whereby an important issue is the stability of that state when various parameters are changed. Such parameters are the external (sensory) signals, the synaptic strengths, which may lead to excitatory or inhibitory interactions, and the various delay times. The lighthouse model had the advantage that all the calculations could be done rather easily. In this way, we could discuss the stability limits and also the effect of pulse shortening. The phase-locked state was established by means of a specific initial condition that does not seem unrealistic. We then studied the kind of stability, whether marginal or asymptotic. We recognized that this feature depends on the kind of perturbations, where we studied those that lead only to marginal stability as well as those that lead to asymptotic stability. The *integrate and fire models*, of which we presented several versions, are mathematically more involved. We focussed our attention on the phase-locked or synchronized state that could be evaluated for all the versions presented. We could also solve the more difficult task of determining the stability that in the present case was of asymptotic nature. This implies that once an initial state is close to a synchronized state it will be fully pulled into that state. We discussed the dependence of the stability on the various parameters that are essentially the same as those of the lighthouse model.

An interesting outcome of the study of all these models was the fact that the requirements for the establishment of a steady synchronized state are rather stringent, in particular the sensory signals must be assumed to be the same. This makes the system highly sensitive to the detection of such coincidences. It may also be a reason why in practice no steady states are observed but only transients. All in all we should bear in mind that in brain functioning we are mostly dealing with episodes rather than with steady states.

The reader may further have noted that there is some dichotomy between synchrony on the one hand and associative memory on the other. In our analysis, the associative memory could be shown to act when we averaged over a number of pulses, i.e. when we looked at *pulse rates*. While for synchronized states we needed the same sensory inputs, in the case of associative memory, at least in general, *different* inputs are needed to obtain, for instance in the visual system, a highly informative image.

Future research may focus on the coexistence or cooperation between synchronized states and those of the associative memory. A first step towards this goal was taken in Sect. 8.12, where we studied the coexistence of synchronized and unsynchronized neuronal groups.

We then indicated what will happen when the neuronal system passes the stability limits. Basically, two different kinds of behavior may occur. The synchronized state is retained, but the pulse intervals are shortened until saturation sets in. In the other case, the synchronized, or, more generally, the phase-locked state is destroyed. Then the nonlinearity inherent in the sigmoid functions S becomes important. While the inclusion of the corresponding effects at the level of spike-trains has not yet been performed, a number of effects were discussed in Sects. 10.1–10.4 in the framework of phase-averaged equations as formulated by Wilson and Cowan. Here spatiotemporal patterns, e.g. stripe patterns, of neural excitations were found. Surely, future work will have to study the combined effects of synchronization and nonlinearity of the sigmoid function, as well as the important role of dendrites.

But let us return to the question of synchronization. Clearly, a central question concerns the biological significance of synchronization. Some scientists, such as Singer, strongly advocate the idea that synchronization is *the* solution to the binding problem that we briefly mentioned at the beginning of this book. Indeed, the experiments on the moving bars hint at a binding or grouping of objects moving in the same direction. We may observe binding at other instances also. For instance, when we wear a blue shirt and a suit with blue stripes, the blue stripes seem to pop up more strongly as compared to the case when our shirt has a different color. Whether this effect has a neuronal underpinning by means of synchronized firing of neurons hasn't been studied yet. More generally speaking, the binding of features and the underlying neuronal processes are parts of a widely unexplored field. Also the interplay between spike-synchronization and associative memory (including learning) is widely unexplored. Progress has been made in the study of synchronization in the case of Parkinson's disease, where Fig. 12.1 shows some more recent remarkable results due to Tass et al. (1999).

In conclusion I may state that some important first steps have been taken towards the exploration of the fascinating phenomenon of spike-synchronization, but also that the field is wide open for further experimental and theoretical research. We must also be aware of the fact that spike-synchronization is just *one* aspect of a wealth of cooperative effects in the neural networks of human and animal brains and that we still need much wider perspectives. To this end, I include a few references for further reading.

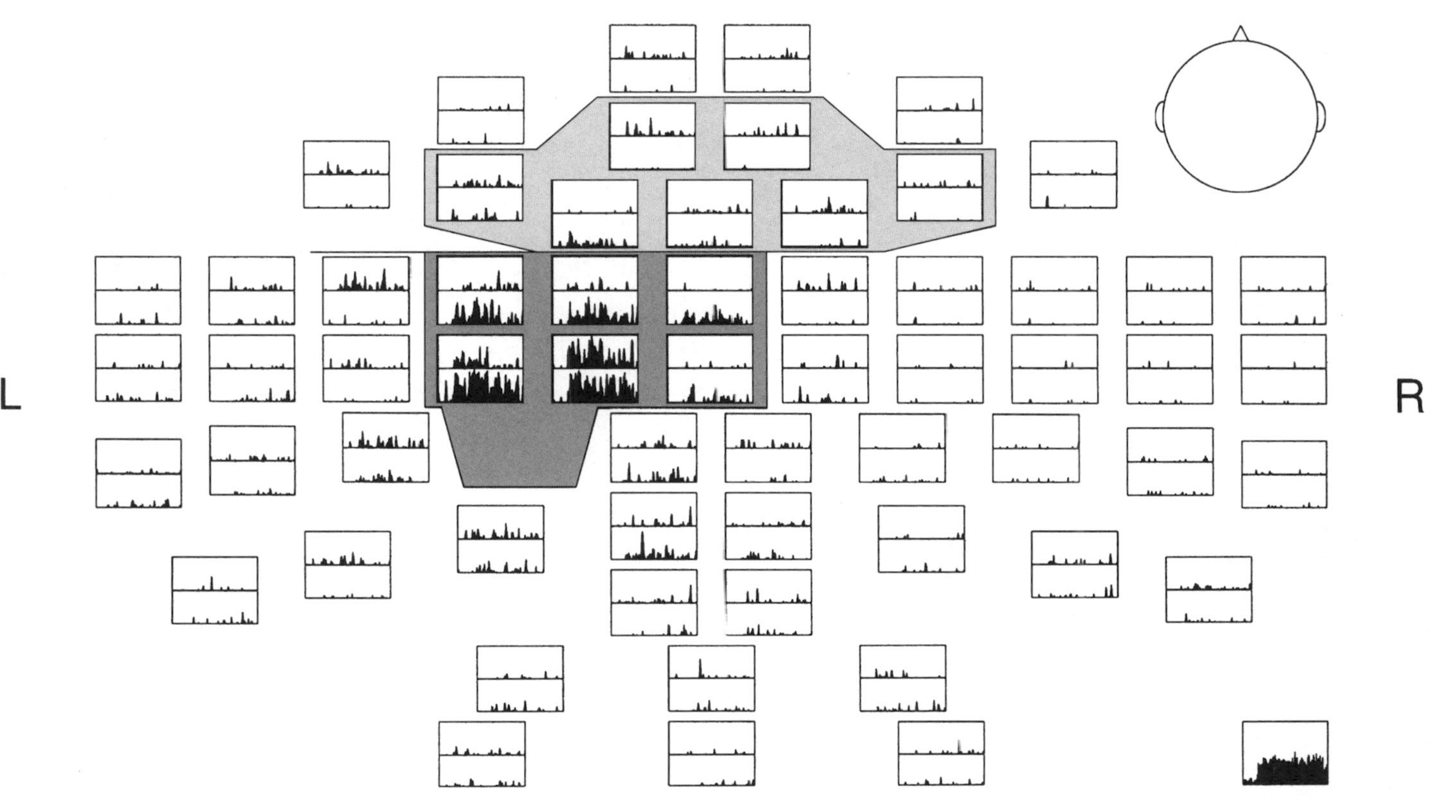

Fig. 12.1. MEG measurements of the cortex during Parkinson tremor. The *shaded areas* exhibit a strong correlation in the sense of phase locking (after Tass et al. (1998))

References

Preface

Haken, H., *Synergetics, an introduction.* 3rd ed., Springer, Berlin 1983
Haken, H., *Principles of brain functioning. A synergetic approach to brain activity, behavior and cognition*, Springer, Berlin 1996
Wilson, H.R., *Spikes, decisions and actions. Dynamical foundations of neuroscience*, Oxford University Press 1999

Introduction

Goal

Epileptic Seizures:

Babloyantz, A., Strange attractors in the dynamics of brain activity. In: *Complex systems – operational approaches.* Haken, H. (ed.), Springer, Berlin 1985
Friedrich, R., Uhl, C.: Synergetic analysis of human electro-encephalograms: Petit-mal epilepsy. In: *Evolution of dynamical structures in complex systems*, R. Friedrich, A. Wunderlin (eds.), Springer, Berlin 1992

Binding Problem:

Singer, W. (1999): Time as coding space? Curr. Opin. Neurobiol. **9**, 189–194
Singer, W. (1999): Neuronal synchrony: a versatile code for the definition of relations? Neuron **24**, 49–65
Singer, W., Gray, C.M. (1995): Visual feature integration and the temporal correlation hypothesis. Ann. Rev. Neurosci. **18**, 555–586
Gray, C.M. (1999): The temporal correlation hypothesis: still alive and well. Neuron **24**, 31–47

Some authors, such as Crick, F. and Koch, C., consider synchronization as a basic mechanism for consciousness; see, for instance:

Wakefield, J. (2001): A mind for consciousness, Sci. American **285**, 26–27

Recent work on Parkinsonian tremor:

Tass, P., *Phase resetting in medicine and biology – stochastic modelling and data analysis.* Springer, Berlin 1999
Tass, P., Rosenblum, M.G., Weule, J., Kurths, J., Pikovsky, A.S., Volkmann, J., Schnitzler, A., Freund, H.J. (1998): Detection of $n : m$ phase locking from noisy data: Application to magnetoencephalography. Phys. Rev. Lett. **81**, 3291–3294

Brain: Structure and Functioning. A Brief Reminder

Adelman, G., Smith, B.H. (eds.), *Elsevier's encyclopedia of neuro-science*, Elsevier Science, Amsterdam 1999

Arbib, M.A. (ed.), *The handbook of brain theory and neural networks.* MIT Press, Cambridge 2000

Bonhoeffer, T., Grinvald, A. (1996): Optical imaging based on intrinsic signals: The methodology. In: Toga, A., Masiotta, J.C. (eds.), *Brain mapping: The methods.* Academic Press, San Diego, CA., 55–97

Grinvald, A., Lieke, E., Frostig, R.D., Gilbert, C.D., Wiesel, T.N. (1986): Functional architecture of cortex revealed by optical imaging of intrinsic signals. Nature **324**, 361–364

Network Models

Bresloff, P.C., Coombes, S. (1998): Phys. Rev. Lett. **81**, 2168, 2384

Ernst, U., Pawelzik, K., Geisel, T. (1995): Synchronization induced by temporal delays in pulse-coupled oscillators. Phys. Rev. Lett. **74**, 1570

Lighthouse model:

Haken, H. (2000): Phase locking in the lighthouse model of a neural net with several delay times. In: Robnik, M., Aizawa, Y., Hasegawa, H., Kuramoto, Y. (eds.). Progr. of Theor. Physics, Supplement **139**, 96–111

Haken, H., Associative memory of a pulse-coupled noisy neural network with delays: The lighthouse model. *Proceedings of traffic and granular flow*, Springer, Berlin 2000

Haken, H., *Laser theory. Encyclopedia of physics XXV/2c*, 2nd ed., Springer, Berlin 1984

Jirsa, V.K., Haken, H. (1996): Field theory of electromagnetic brain activity, Phys. Rev. Lett. **77**, 960–963; (1996): Derivation of a field equation of brain activity, Journal of Biol. Physics **22**, 101–112

Kelso, J.A.S. (1990): Phase transitions: Foundations of behavior. In: Haken, H., Stadler, M. (eds.), *Synergetics of cognition*, 249–295, Springer, Berlin

Kelso, J.A.S., *Dynamic patterns: The self-organization of brain and behavior*, MIT Press, Boston 1995

Kuramoto, Y., *Chemical oscillations, waves, and turbulence*, Springer, Berlin 1984

Kuramoto, Y., Nishikawa, I. (1987): J.Stat. Phys. **49**, 569

Kuramoto, Y. (1991): Physica D **50**, 15

Mirollo, R.E., Strogatz, S.H. (1990): Synchronization of pulse-coupled biological oscillators. SIAM J. Appl. Math. **50**, 1645–1662

Nunez, P.L. (1974): The brain wave equation: A model for the EEG, Mathematical Biosciences **21**, 279–297

Nunez, P.L., *Electric fields of the brain*, Oxford University Press 1981

Nunez, P.L., *Neocortical dynamics and human EEG rhythms*, Oxford University Press 1995

Peskin, C.S. (1975): Mathematical aspects of heart physiology, Courant Institute of Mathematical Sciences, New York University, New York, 268–278

For reviews on synchronization mechanisms in the brain cf.:

Strogatz, S.M. (2000): From Kuramoto to Crawford: exploring the onset of synchronization in populations of coupled oscillators, Physica D **143**
Sturz, A.K., König, P. (2001): Mechanisms for synchronization of neuronal activity, Biol. Cybern. **84**, 153

Viterbi, A.J., *Principles of coherent communication*, McGraw Hill, New York 1966
Wilson, H.R. and Cowan, J.D. (1972): Excitatory and inhibitory interactions in localized populations of model neurons, Biophysical Journal, **12**, 1–24
Winfree, A.T. (1967): J. Theor. Biol. **16**, 15
Winfree, A.T., *The Geometry of biological time*, Springer, Berlin; 2nd ed. to be published 2001

For a general survey on phase synchronization cf. the papers in the volumes 10, Ns. 10 and 11 (2000) in the Int. J. of Bifurcation and Chaos, on phase synchronization and its applications. Guest editor: Jürgen Kurths

In the context of our book, the contribution by Freeman is of particular interest:

Freeman, W.J. (2000): Characteristics of the synchronization of brain activity imposed by finite conduction velocities of axons, Int. J. Bifurcation and Chaos **10**, 2307–2322

The Neuron – Building Block of the Brain

Structure and Basic Functions

Bullock, T.H., Orkland, R., Grinnell, A., *Introduction to nervous systems*, Freeman, W.H. and Co., San Francisco 1977
Dowling, J.E., *Neurons and networks. An introduction to neuroscience.* Harvard University Press, Cambridge 1992
Kandel, E.R., Schwartz, J.H., Jessel, T.M., *Principles of neural sciences.* 4th ed., McGraw Hill, N.Y. 2000

For the more recent method of optical imaging cf.:

Malonek, D., Grinvald, A. (1996): Interactions between electrical activity and cortical microcirculation revealed by imaging spectroscopy: implications for functional brain mapping. Science **272**, 551–554

Synapses – The Local Contacts

For a recent detailed study cf.

Llinas, R.R., *The squid giant synapse. A model for chemical transmission*, University Press, Oxford 1999

Naka–Rushton Relation

Albrecht, D.G., Hamilton, D.B. (1982): Striate cortex of monkey and cat: contrast response function. J. Neurophys. **48**, 217–237
Naka, K.I., Rushton, W.A. (1966): S-potentials from colour units in the retina of fish. J. Physiol. **185**, 584–599

Learning and Memory

Hebb, D.O., *The organization of behavior*, Wiley, New York 1949

Neuronal Cooperativity

Abeles, M., *Local cortical circuits. An electrophysiological study.* Springer, Berlin 1982

Hubel, D.H., Wiesel, T.N. (1962): Receptive fields, binocular interaction and functional architecture in the cat's visual cortex. J. Physiol. (London) **160**, 106–154

Hubel, D.H., Wiesel, T.N. (1977): Functional architecture of macaque monkey visual cortex. Proc. R. Soc. Lond. B. **198**, 1–59

Hubel, D.H. (1982): Cortical neurobiology: a slonted historical perspective. Ann. Rev. Neurosci. **5**, 363–370

Hubel, D.H., *Eye, brain and vision*, Freeman, W.J., New York 1995

Interlude: A Minicourse on Correlations

Komatsu, Y., Nakajima, S., Toyama, K., Fetz, E. (1988): Intracortical connectivity revealed by spike-triggered averaging in slice preparations of cat visual cortex. Brain Res. **442**, 359–362

Krüger, J. (ed.), *Neuronal cooperativity*, Springer, Berlin 1991

Toyama, K. (1991): The structure-function problem in visual cortical circuitry studied by cross-correlation techniques and multi-channel recording, 5–29, in: Krüger, J., l.c.

Mesoscopic Neuronal Cooperativity

Eckhorn, R., Bauer, R., Jordan, W., Brosch, M., Kruse W., Munk, M., Reitboeck, H.J. (1988): Coherent oscillations: a mechanism of feature linking in the visual cortex? Multiple electrode and correlation analyses in the cat. Biol. Cybern. **60**, 121–130

Eckhorn, R., Reitboeck, H.J., Arndt, M., Dicke, P. (1990): Feature linking among distributed assemblies: Simulations and results from cat visual cortex. Neural Computation **2**, 293–306

Eckhorn, R. (1991): Stimulus-specific synchronization in the visual cortex: linking of local features into global figures? 184–224, in: Krüger, J., see below

Eckhorn, R., Principles of global visual processing of local features can be investigated with parallel single-cell and group-recordings from the visual cortex. In: *Information processing in the cortex, experiments and theory*, Aertsen, A., Braitenberg, V. (eds.), Springer, Berlin 1992

Eckhorn, R., Obermueller, A. (1993): Single neurons are differently involved in stimulus-specific oscillations in cat visual cortex. Experim. Brain Res. **95**, 177–182

Eckhorn, R. (1994): Oscillatory and non-oscillatory synchronizations in the visual cortex and their possible roles in associations of visual features. Prog. Brain Res. **102**, 404–426

Eckhorn, R., Cortical processing by fast synchronization: high frequency rhythmic and non-rhythmic signals in the visual cortex point to general principles of spatiotemporal coding. In: *Time and Brain*, Miller, R. (ed.), Gordon and Breach, Lausanne, Switzerland 2000

Engel, A.K., König, P., Singer, W. (1991): Direct physiological evidence for scene segmentation by temporal coding. Proc. Natl. Acad. Sci., USA **88**, 9136–9140

Engel, A.K., König, P., Schillen, T.B. (1992): Why does the cortex oscillate? Curr. Biol. **2**, 332–334

Freeman, W.J. (1979a): EEG analysis gives model of neuronal template-matching mechanism for sensory search with olfactory bulb. Biol. Cybern. **35**, 221–234

Freeman, W.J. (1979b): Nonlinear dynamics of paleocortex manifested in the olfactory EEG. Biol. Cybern. **35**, 21–37

Gray, C.M., Singer, W. (1987): Stimulus-dependent neuronal oscillations in the cat visual cortex area 17. IBRO Abstr. Neurosci. Lett. Suppl. **22**, 1301

Gray, C.M., Singer, W. (1989): Stimulus-specific neuronal oscillations in orientation columns of cat visual cortex. Proc. Natl. Acad. Sci., USA **86**, 1698–1702

Gray, C.M., König, P., Engel, A.K., Singer, W. (1989): Oscillatory responses in cat visual cortex exhibit inter-columnar synchronization which reflects global stimulus properties. Nature **338**, 334–337

Gray, C.M., Engel, A.K., König, P., Singer, W. (1990): Stimulus-dependent neuronal oscillations in cat visual cortex: receptive field properties and feature dependence. Eur. J. Neurosci. **2**, 607–619

Gray, C.M., Maldonado, P.E., Wilson, M., McNaughton, B. (1995): Tetrodes markedly improve the reliability and yield of multiple single-unit isolation from multi-unit recordings in cat striate cortex. J. Neurosci. Methods **63**, 43–54

Jürgens, E., Eckhorn, R. (1997): Parallel processing by a homogeneous group of coupled model neurons can enhance, reduce and generate signal correlations. Biol. Cybern. **76**, 217–227

König, P., Schillen, T.B. (1991): Stimulus-dependent assembly formation of oscillatory responses. I. Synchronization. Neural Comput. **3**, 155–166

Krüger, J. (ed.), *Neuronal cooperativity*, Springer, Berlin 1991

Legatt, A.D., Arezzo, J., Vaugham, Jr., H.G. (1980): Averaged multiple unit activity as an estimate of phasic changes in local neuronal activity. J. Neurosci. Methods **2**, 203–217

Mitzdorf, U. (1987): Properties of the evoked potential generators: current source density analysis of visually evoked potentials in the cat cortex. Int. J. Neurosci. **33**, 33–59

Singer, W. (1991): The formation of cooperative cell assemblies in the visual cortex, 165–183, in: Krüger, J., l.c.

Spikes, Phases, Noise: How to Describe Them Mathematically? We Learn a Few Tricks and Some Important Concepts

Kicks – Random Kicks or a Look at Soccer Games

For this kind of approach cf.

Haken, H. (1983): *Synergetics, an introduction*, 3rd. ed., Springer, Berlin

Schottky, W. (1918): Ann. Phys. (Leipzig) **57**, 541 (studied this kind of "shot noise".)

Noise Is Inevitable. Brownian Motion and Langevin Equation

Einstein, A. (1905): Ann. Physik (Leipzig) **17**, 549
Gardiner, C.W., *Handbook of stochastic methods for physics, chemistry and the natural sciences*, 4th print, Springer, Berlin 1997
Haken, H. (1983): l.c. (sect. 4.4)
Langevin, P. (1908): Comptes rendue **146**, 530

Noise in Active Systems

For a comprehensive survey see

Haken, H., *Laser theory*, Springer, Berlin 1970

The Concept of Phase

How To Determine Phases From Experimental Data? Hilbert-Transform

Using the Hilbert-transform, the "analytic signal" (4.152), (4.153) was constructed by

Gabor, D. (1946): Theory of communication, J.IEE London **93** (3), 429–457

Phase Noise

Origin of Phase Noise

see e.g.

Haken, H., *Laser theory*, Springer, Berlin 1970

The Lighthouse Model. Two Coupled Neurons

Haken, H. (1999): What can synergetics contribute to the understanding of brain functioning?. In: Uhl, C. (ed.) (1999): *Analysis of neuro-physiological brain functioning*, 7–40, Springer, Berlin
Haken, H. (2000): Quasi-discrete dynamics of a neural net: the lighthouse model. Discrete Dynamics in Nature and Society **4**, 187–200
Haken, H. (2000): In: Broomhead, D.S., Luchinskaya, E.A., Mc.Clintock, P.V.E., Mullin, T. (eds.), *Stochaos, stochastic and chaotic dynamics in the lakes*. American Institute of Physics, Woodbury, N.Y.

The Lighthouse Model. Many Coupled Neurons

Haken, H. (2000): Phase locking in the lighthouse model of a neural net with several delay times, in: Robnik, M., Aizawa, Y., Hasegawa, H., Kuramoto, Y., eds., Progr. Theor. Physics Supplement **139**, 96–111
Haken, H. (2000): Effect of delay on phase locking in a pulse coupled neural network, Eur. Phys. J. B **18**, 545–550
Haken, H. (2000): Quasi-discrete dynamics of a neural net: the lighthouse model. Discrete Dynamics in Nature and Society **4**, 187–200

Integrate and Fire Models (IFM)

The General Equations of IFM

These models have been treated by a number of authors. See also Sect. 7.2 and Chap. 8:

Kuramoto, Y. (1991): Physica D **50**, 15
Abbott, L.F., van Vreeswijk, C. (1993): Phys. Rev. E **48**, 1483
Treves, A. (1993): Network **4**, 256
Gerstner, W., van Hemmen, J.L. (1993): Phys. Rev. Lett **71**, 312
Gerstner, W. (1995): Phys. Rev. E **51**, 738

There are numerous further publications on the theory of spiking neurons by W. Gerstner and L.J. van Hemmen, and we refer the reader to the home pages of these authors.

Peskin's Model

Peskin, C.S. (1975): Mathematical aspects of heart physiology, Courant Institute of Mathematical Sciences, New York University, New York, 268–278
Mirollo, R.E., Strogatz, S.H. (1990): Synchronization of pulse-coupled biological oscillators. SIAM J. Appl. Math. **50**, 1645–1662

Many Neurons, General Case, Connection with Integrate and Fire Model

The results of this chapter are based on unpublished work by the present author.

The Phase-Locked State
Stability of the Phase-Locked State: Eigenvalue Equations

For an alternative approach cf.:

Bressloff, P.C., Coombes, S. (1998): Phys. Rev. Lett. **81**, 2168
Bressloff, P.C., Coombes, S. (1998): Phys. Rev. Lett. **81**, 2384

For earlier work in which the firing times are considered as the fundamental dynamical variables, cf.

Keener, J.P., Hoppenstadt, F.C., Rinzel, J. (1981): SIAM J. Appl. Math. **41**, 503
van Vreeswijk, C. (1996): Phys. Rev. E **54**, 5522

Some extensions of integrate and fire models with mean field coupling:

Impact of periodic forcing:

Coombes, S., Bressloff, P.C. (1999): Mode locking and Arnold tongues in integrate and fire neural oscillators, Phys. Rev. E **60**, 2086–2096

Impact of dynamic synapses

Bressloff, P.C. (1999): Mean-field theory of globally coupled integrate-and-fire oscillators with dynamic synapses. Phys. Rev. E **60**, 2160–2170

Details concerning the dendritic tree

Bressloff, P.C. (1996): New mechanism for neural pattern formation, Phys. Rev. Lett. **76**, 4644–4647
Bressloff, P.C., Coombes, S. (1997): Physics of the extended neuron. Int. J. Mod. Phys. B. **11**, 2343–2392 contains many further references

Phase Locking via Sinusoidal Couplings

Coupling Between Two Neurons

Phase oscillator concept of neurons:

Cohen, A.H., Holmes, P.J., Rand, R.H. (1982): The nature of the coupling between segmental oscillators of the lamprey spinal generator for locomotion: a mathematical model. J. Math. Biology **13**, 345–369

Kuramoto, Y., *Chemical oscillations, waves and turbulence*, Springer, Berlin 1984

Radio-engineering:

Viterbi, A.J., *Principles of coherent communication*, McGraw Hill, N.Y. 1966

Laser-physics:

Haken, H., *Laser theory*, Springer, Berlin 1970

A Chain of Coupled Phase Oscillators

Cohen, A.H., Ermentrout, G.B., Kiemel, T., Kopell, N., Sigvardt, K.A., and Williams, T.L. (1992): Modelling of intersegmental coordination in the lamprey central pattern generator for locomotion. TINS **15**, 434–438
Gray, J., *Animal locomotion*. Weidenfield and Nicolson, London 1968
Grillner, S. (1996): Neural networks for vertebrate locomotion. Sci. Am. **274**, 64–69
Grillner, S., Deliagina, T., Ekeberg, Ö., El Manira, A., Hill, R.H. (1995): Neural networks that coordinate locomotion in lamprey. TINS **18**, 270–279
Wilson, H.R., *Spikes, decisions and actions*, Oxford University Press 1999

Coupled Finger Movements

Haken, H. (1983): Synopsis and introduction, in Başar, E., Flohr, H., Haken, H., Mandell, A.J. (eds.): *Synergetics of the brain*, Springer, Berlin, Heidelberg, New York, 3–25
Kelso, J.A.S. (1984): Phase transitions and critical behavior in human bimanual coordination, American Journal of Physiology: Regulatory, Integrative and Comparative Physiology **15**, R 1000–R 1004
Haken, H., Kelso, J.A.S., Bunz, H. (1985): A theoretical model of phase transitions in human movement, Biol. Cybern. **53**, 247–257

Schöner, G., Haken, H., Kelso, J.A.S. (1986): A stochastic theory of phase transitions in human movement, Biol. Cybern. **53**, 247–257

Quadruped Motion

Schöner, G., Yiang, W.Y., Kelso, J.A.S. (1990): A synergetic theory of quadropedal gaits and gait transitions, J. Theor. Biol. **142**, 359–391
Lorenz, W. (1987): Nichtgleichgewichtsphasenübergänge bei Bewegungskoordination, Diplom Thesis, University of Stuttgart
Collins, J.J., Stewart, I.N. (1993): Coupled nonlinear oscillators and the symmetries of animal gaits, J. Nonlinear Sci. **3**, 349–392
Hoyt, D.F., Taylor, C.R. (1981): Gait and the energetics of locomotion in horses, Nature **292**, 239

Populations of Neural Phase Oscillators

Tass, P., Haken, H. (1996): Synchronization in networks of limit cycle oscillators. Z. Phys. B **100**, 303–320
Tass, P., Haken, H. (1996): Synchronized oscillations in the visual cortex – a synergetic model. Biol. Cybern. **74**, 31–39
Tass, P. (1997): Phase and frequency shifts in a population of phase oscillators. Phys. Rev. E **56**, 2043–2060
Tass, P., *Phase resetting in medicine and biology – stochastic modelling and data analysis*. Springer, Berlin 1999

Tass, P. (1996): Resetting biological oscillators – a stochastic approach. J. Biol. Phys. **22**, 27–64
Tass, P. (1996): Phase resetting associated with changes of burst shape. J. Biol. Phys. **22**, 125–155
Tass, P. (2001): Effective desynchronization by means of double-pulse phase resetting. Europhys. Lett. **53**, (1), 15–21

Pulse-Averaged Equations

Survey

Nunez, P.L., *Neocortical dynamics and human EEG rhythms*, Oxford University Press, New York, Oxford (with further contributions by Cutillo, B.A., Gevins, A.S., Ingber, L., Lopes da Silva, F.H., Pilgreen, K.L., Silberstein, R.B.) 1985

The Wilson-Cowan Equations

Wilson, H.R., Cowan, J.D. (1972): Excitatory and inhibitory interactions in localized populations of model neurons. Biophysical Journal **12**, 1–24
Wilson, H.R., Cowan, J.D. (1973): A mathematical theory of the functional dynamics of cortical and thalamic nervous tissue. Kybernetik **13**, 55–80
Wilson, H.R., *Spikes, decisions and actions*, Oxford University Press 1999

Cortical Dynamics Described by Wilson-Cowan Equations

Ermentrout, B. (1998): Neural networks as spatio-temporal pattern-forming systems. Rep. Prog. Phys. **61**, 353–430
Wilson, H.R., Cowan, J.D. (1973): A mathematical theory of the functional dynamics of cortical and thalamic nervous tissue. Kybernetik **13**, 55–80

Visual Hallucinations

Ermentrout, G.B. and Cowan, J.D. (1979): A mathematical theory of visual hallucination patterns. Biol. Cybernetics **34**, 137–150
Tass, P. (1995): Cortical pattern formation during visual hallucinations. J. Biol. Phys. **21**, 177–210
Tass, P. (1997): Oscillatory cortical activity during visual hallucinations. J. Biol. Phys. **23**, 21–66

Jirsa-Haken-Nunez Equations

Jirsa, V.K., Haken, H. (1997): A derivation of a macroscopic field theory of the brain from the quasi-microscopic neural dynamics. Physica D **99**, 503–516
Nunez, P.L. (1974): The brain wave equation: A model for the EEG, Mathematical Biosciences **21**, 279–297
Nunez, P.L., *Electric fields of the brain*, Oxford University Press 1981
Nunez, P.L., *Neocortical dynamics and human EEG rhythms*, Oxford University Press 1995

An Application to Movement Control

Jirsa, V.K., Haken, H. (1997): l.c.
Haken, H., *Principles of brain functioning*, Springer, Berlin 1996
Kelso, J.A.S., *Dynamic patterns: The self-organization of brain and behavior*, MIT Press, Boston 1995

Experiments:

Kelso, J.A.S., Bressler, S.L., Buchanan, S., DeGuzman, G.C., Ding, M., Fuchs, A., Holroyd, T. (1991): Cooperative and critical phenomena in the human brain revealed by multiple SQUID's. In: *Measuring chaos in the human brain*, Duke, D., Pritchard, W. (eds.), World Scientific, Singapore
Kelso, J.A.S., Bressler, S.L., Buchanan, S., DeGuzman, G.C., Ding, M., Fuchs, A., Holroyd, T. (1992): A phase transition in human brain and behavior, Physics Letters A **169**, 134–144

Theory:

Friedrich, R., Fuchs, A., Haken, H. (1992): Spatio-temporal EEG-patterns. In: *Rhythms in biological systems*, Haken, H., Köpchen, H.P. (eds.), Springer, Berlin
Fuchs, A., Kelso, J.A.S., Haken, H. (1992): Phase transitions in the human brain: Spatial mode dynamics, International Journal of Bifurcation and Chaos **2**, 917–939
Jirsa, V.K., Friedrich, R., Haken, H., Kelso, J.A.S. (1994): A theoretical model of phase transitions in the human brain, Biol. Cybern. **71**, 27–35
Jirsa, V.K., Friedrich, R., Haken, H.: Reconstruction of the spatio-temporal dynamics of a human magnetoencephalogram (unpublished)

The Single Neuron

Hodgkin-Huxley Equations

Hodgkin, A.L., Huxley, A.F. (1952): A quantitative description of membrane current and its application to conduction and excitation in nerve. J. Physiol. **117**, 500–544
Rinzel, J. (1985): Excitation dynamics: insights from simplified membrane models. Fed. Proc. **44**, 2944–2946
Wilson, H.R., *Spikes, decisions and actions*, Oxford University Press 1999

FitzHugh-Nagumo Equations

FitzHugh, R. (1961): Impulses and physiological states in models of nerve membrane. Biophys. J. **1**, 445–466
Nagumo, J.S., Arimoto, S., Yoshizawa, S. (1962): An active pulse transmission line simulating a nerve axon. Proc. IRE **50**, 2061–2070
Wilson, H.R., *Spikes, decisions and actions*, Oxford University Press 1999

Some Generalization of the Hodgkin-Huxley Equations

Connor, J.A., Walter, D., and McKown, R. (1977): Neural repetitive firing: modifications of the Hodgkin-Huxley axon suggested by experimental results from crustacean axons. Biophys. J. **18**, 81–102
Wilson, H.R. (1999): Simplified dynamics of human and mammalian neocortical neurons, J. Theor. Biol. **200**, 375–388
Wilson, H.R., *Spikes, decisions and actions*, Oxford University Press 1999

Dynamical Classes of Neurons

Chay, T.R. and Heizer, J. (1983): Minimal model for membrane oscillations in the pancreatic beta-cell. Biophys. J. **42**, 181–189
Connors, B.W. and Gutnick, M.J. (1990): Intrinsic firing patterns of diverse neocortical neurons. TINS **13**, 99–104
Gray, C.M. and McCormick, D.A. (1996): Chattering cells: superficial pyramidal neurons contributing to the generation of synchronous oscillations in the visual cortex. Science **274**, 109–113
Gutnick, M.J. and Crill, W.E. (1995): The cortical neuron as an electrophysiological unit. In: *The cortical neuron*, ed. Gutnick, M.J. and Moody, I., 33–51, Oxford University Press, New York
Wilson, H.R. (1999): Simplified dynamics of human and mammalian neocortical neurons, J. Theor. Biol. **200**, 375–388
Wilson, H.R., *Spikes, decisions and actions*, Oxford University Press 1999

Some Conclusions on Network Models

Wilson, H.R., *Spikes, decisions and actions*, Oxford University Press 1999

Conclusion and Outlook

Study of Parkinsonian tremor:

Tass, P., Rosenblum, M.G., Weule, J., Kurths, J., Pikovsky, A., Volkmann, J., Schnitzler, A., Freund, H.J. (1998): Detection of $n : m$ phase locking from noisy data: Application to magnetoencephalography. Phys. Rev. Lett **81**, 3291–3294

Wider perspectives:

Crick, F. (1984): Function of the thalamic reticular complex: The searchlight hypothesis. Proc. Natl. Acad. Sci. USA **81**, 4586–4590

Freeman, W.J., *Societies of brain*, Lawrence Erlbaum Associates, Mahwah, N.Y. 1995

Freeman, W.J., *How brains make up their minds?* Columbia University Press 1999

Haken, H., *Principles of brain functioning*, Springer, Berlin 1996

Pikovsky, A., Rosenblum, M., Kurths, J. (2000): Phase synchronization in regular and chaotic systems, Int. J. Bifurcation and Chaos **10**, 2291–2305

Scott, A.C., *Stairway to the mind*, Copernicus-Springer, N.Y. 1995

Index

Springer Series in Synergetics

Synergetics An Introduction 3rd Edition
By H. Haken

Synergetics A Workshop
Editor: H. Haken

Synergetics Far from Equilibrium
Editors: A. Pacault, C. Vidal

Structural Stability in Physics
Editors: W. Güttinger, H. Eikemeier

Pattern Formation by Dynamic Systems and Pattern Recognition
Editor: H. Haken

Dynamics of Synergetic Systems
Editor: H. Haken

Problems of Biological Physics
By L. A. Blumenfeld

Stochastic Nonlinear Systems in Physics, Chemistry, and Biology
Editors: L. Arnold, R. Lefever

Numerical Methods in the Study of Critical Phenomena
Editors: J. Della Dora, J. Demongeot, B. Lacolle

The Kinetic Theory of Electromagnetic Processes By Yu. L. Klimontovich

Chaos and Order in Nature
Editor: H. Haken

Nonlinear Phenomena in Chemical Dynamics Editors: C. Vidal, A. Pacault

Handbook of Stochastic Methods for Physics, Chemistry, and the Natural Sciences 2nd Edition
By C. W. Gardiner

Concepts and Models of a Quantitative Sociology The Dynamics of Interacting Populations By W. Weidlich, G. Haag

Noise-Induced Transitions Theory and Applications in Physics, Chemistry, and Biology By W. Horsthemke, R. Lefever

Physics of Bioenergetic Processes
By L. A. Blumenfeld

Evolution of Order and Chaos in Physics, Chemistry, and Biology
Editor: H. Haken

The Fokker-Planck Equation
2nd Edition By H. Risken

Chemical Oscillations, Waves, and Turbulence By Y. Kuramoto

Advanced Synergetics
2nd Edition By H. Haken

Stochastic Phenomena and Chaotic Behaviour in Complex Systems
Editor: P. Schuster

Synergetics – From Microscopic to Macroscopic Order Editor: E. Frehland

Synergetics of the Brain
Editors: E. Başar, H. Flohr, H. Haken, A. J. Mandell

Chaos and Statistical Methods
Editor: Y. Kuramoto

Dynamics of Hierarchical Systems
An Evolutionary Approach
By J. S. Nicolis

Self-Organization and Management of Social Systems Editors: H. Ulrich, G. J. B. Probst

Non-Equilibrium Dynamics in Chemical Systems
Editors: C. Vidal, A. Pacault

Self-Organization Autowaves and Structures Far from Equilibrium
Editor: V. I. Krinsky

Temporal Order Editors: L. Rensing, N. I. Jaeger

Dynamical Problems in Soliton Systems
Editor: S. Takeno

Complex Systems – Operational Approaches in Neurobiology, Physics, and Computers Editor: H. Haken

Dimensions and Entropies in Chaotic Systems Quantification of Complex Behavior 2nd Corr. Printing
Editor: G. Mayer-Kress

Selforganization by Nonlinear Irreversible Processes
Editors: W. Ebeling, H. Ulbricht

Instabilities and Chaos in Quantum Optics
Editors: F. T. Arecchi, R. G. Harrison

Nonequilibrium Phase Transitions in Semiconductors Self-Organization Induced by Generation and Recombination Processes By E. Schöll

Temporal Disorder in Human Oscillatory Systems
Editors: L. Rensing, U. an der Heiden, M. C. Mackey

Springer Series in Synergetics

The Physics of Structure Formation Theory and Simulation Editors: W. Guttinger, G. Dangelmayr

Computational Systems – Natural and Artificial Editor: H. Haken

From Chemical to Biological Organization Editors: M. Markus, S. C. Müller, G. Nicolis

Information and Self-Organization A Macroscopic Approach to Complex Systems 2nd Edition By H. Haken

Propagation in Systems Far from Equilibrium Editors: J. E. Wesfreid, H. R. Brand, P. Manneville, G. Albinet, N. Boccara

Neural and Synergetic Computers Editor: H. Haken

Cooperative Dynamics in Complex Physical Systems Editor: H. Takayama

Optimal Structures in Heterogeneous Reaction Systems Editor: P. J. Plath

Synergetics of Cognition Editors: H. Haken, M. Stadler

Theories of Immune Networks Editors: H. Atlan, I. R. Cohen

Relative Information Theories and Applications By G. Jumarie

Dissipative Structures in Transport Processes and Combustion Editor: D. Meinköhn

Neuronal Cooperativity Editor: J. Krüger

Synergetic Computers and Cognition A Top-Down Approach to Neural Nets By H. Haken

Foundations of Synergetics I Distributed Active Systems 2nd Edition By A. S. Mikhailov

Foundations of Synergetics II Complex Patterns 2nd Edition By A. S. Mikhailov, A. Yu. Loskutov

Synergetic Economics By W.-B. Zhang

Quantum Signatures of Chaos 2nd Edition By F. Haake

Rhythms in Physiological Systems Editors: H. Haken, H. P. Koepchen

Rhythms in Physiological Systems Editors: H. Haken, H. P. Koepchen

Quantum Noise 2nd Edition By C. W. Gardiner, P. Zoller

Nonlinear Nonequilibrium Thermodynamics I Linear and Nonlinear Fluctuation-Dissipation Theorems By R. Stratonovich

Self-organization and Clinical Psychology Empirical Approaches to Synergetics in Psychology Editors: W. Tschacher, G. Schiepek, E. J. Brunner

Nonlinear Nonequilibrium Thermodynamics II Advanced Theory By R. Stratonovich

Limits of Predictability Editor: Yu. A. Kravtsov

On Self-Organization An Interdisciplinary Search for a Unifying Principle Editors: R. K. Mishra, D. Maaß, E. Zwierlein

Interdisciplinary Approaches to Nonlinear Complex Systems Editors: H. Haken, A. Mikhailov

Inside Versus Outside Endo- and Exo-Concepts of Observation and Knowledge in Physics, Philosophy and Cognitive Science Editors: H. Atmanspacher, G. J. Dalenoort

Ambiguity in Mind and Nature Multistable Cognitive Phenomena Editors: P. Kruse, M. Stadler

Modelling the Dynamics of Biological Systems Editors: E. Mosekilde, O. G. Mouritsen

Self-Organization in Optical Systems and Applications in Information Technology 2nd Edition Editors: M.A. Vorontsov, W. B. Miller

Principles of Brain Functioning A Synergetic Approach to Brain Activity, Behavior and Cognition By H. Haken

Synergetics of Measurement, Prediction and Control By I. Grabec, W. Sachse

Predictability of Complex Dynamical Systems By Yu. A. Kravtsov, J. B. Kadtke

Springer Series in Synergetics

Interfacial Wave Theory of Pattern Formation
Selection of Dentritic Growth and Viscous Fingerings in Hele–Shaw Flow By Jian-Jun Xu

Asymptotic Approaches in Nonlinear Dynamics
New Trends and Applications
By J. Awrejcewicz, I. V. Andrianov, L. I. Manevitch

Brain Function and Oscillations
Volume I: Brain Oscillations. Principles and Approaches
Volume II: Integrative Brain Function. Neurophysiology and Cognitive Processes
By E. Başar

Asymptotic Methods for the Fokker–Planck Equation and the Exit Problem in Applications
By J. Grasman, O. A. van Herwaarden

Analysis of Neurophysiological Brain Functioning Editor: Ch. Uhl

Phase Resetting in Medicine and Biology
Stochastic Modelling and Data Analysis
By P. A. Tass

Self-Organization and the City By J. Portugali

Critical Phenomena in Natural Sciences
Chaos, Fractals, Selforganization and Disorder: Concepts and Tools By D. Sornette

Spatial Hysteresis and Optical Patterns
By N. N. Rosanov

Nonlinear Dynamics of Chaotic and Stochastic Systems
Tutorial and Modern Developments
By V. S. Anishchenko, V. V. Astakhov, A. B. Neiman, T. E. Vadivasova, L. Schimansky-Geier

Synergetic Phenomena in Active Lattices
Patterns, Waves, Solitons, Chaos
By V. I. Nekorkin, M. G. Velarde

Brain Dynamics
Synchronization and Activity Patterns in Pulse-Coupled Neural Nets with Delays and Noise By H. Haken

From Cells to Societies
Models of Complex Coherent Action
By A. S. Mikhailov, V. Calenbuhr

Printing (Computer to Film): Saladruck Berlin
Binding: Stürtz AG, Würzburg